OIL TERMS

A Dictionary of Terms used in
Oil Exploration and Development

OIL TERMS

A Dictionary of Terms used in
Oil Exploration and Development

Leo Crook

INTERNATIONAL PUBLICATIONS SERVICE
New York

Published in 1976 by
International Publications Service
114 East 32nd Street
New York, N.Y. 10016

Library of Congress Cataloging in Publication Data

Crook, Leo
Oil Terms

1. Oil well drilling—Dictionaries.
2. Prospecting—Dictionaries.
3. Petroleum engineering—Dictionaries.
I. Title

TN865.C76 622'.33'803 75–38510
ISBN 0–8002–0164–7

Printed and bound in Great Britain

CONTENTS

(Continued)

LIST OF ILLUSTRATIONS

List of illustrations (continued)

List of illustrations (continued)

ACKNOWLEDGEMENTS

I record my sincere thanks to all who have contributed their knowledge and assistance in the compilation of this book. In particular, I must thank many organisations operating in oil exploration and development who have given permission for illustrations to be reproduced and statistics and other information to be reprinted.

American Petroleum Institute, for the use of illustrations from *Primer of Oil and Gas Production.*

B J Servie B.V. for illustrations of some of their equipment.

The British Petroleum Co. Ltd, for the use of illustrations from *Our Industry, Petroleum.*

Cameron Ironworks Ltd for diagrams from their *Oil Tool Products* catalogue.

Christensen Diamond Products (UK) Ltd, for the use of illustrations from their publications.

Dresser Europe S.A. for reproductions from their *World Oil* composite catalogue.

Eastman Oil Well Survey Co, Houston, Texas, for reproductions of their equipment.

Eastman Whipstock (UK) Ltd for use of their directional drilling illustration.

Hughes Tool Company, Houston, Texas, for reproductions of illustrations and terminology from the Hughes Catalog.

The Institute of Petroleum for statistics reproduced in the appendices.

Lockheed Petroleum Services Ltd.

Lufkin Industries (UK) Ltd for a reproduction from their pumping units catalogue.

Noroil, Stavanger, Norway, for use of illustrations from *Noroil.*

The University of Oklahoma Press for the use of reproductions from *Energy under the Oceans.*

Oweco Ltd for the use of illustrations of their products.

Schlumberger Inland Services, Inc. for the use of illustrations from their published literature.

The Universitv of Texas at Austin for the use of illustrations from *Primer of Oil Well Drilling.*

All items reproduced on the following pages from published material have, wherever possible, been acknowledged to the copyright owners, but the author would be grateful for notification of any errors or omissions so that they may be rectified in the next edition.

Leo Crook

INTRODUCTION

Much of the equipment used in the oil industry and especially in the exploration and drilling operations is quite unique when compared with other fields of engineering. Over the years this new technology has developed its own terminology and it is of great value, if not essential, that newcomers to the industry, journalists, and those wishing to compete with longstanding suppliers should familiarise themselves with every possible aspect of the exploration, drilling and production methods, equipment, and the 'language' involved.

In the following pages, therefore, this terminology is listed and described and much of the equipment peculiar to the exploration and drilling operations is illustrated. This book will be a valuable reference work, even for the experienced oil man but, more importantly, it will be useful as an introduction for the many individuals, companies and organisations, who whilst experienced in their own industries, have little or no previous contact with the exciting field of oil exploration, drilling and production.

OIL EXPLORATION AND DEVELOPMENT

The presence of promising reservoir formations in the seas around the British coastline have been known to geologists for many years but only recently have engineers been in a position to tackle the multitude of almost insurmountable problems presented by water depths and weather conditions never before encountered anywhere in the world. Although engineers are at present only considering production problems in a water depth of 450 to 600 feet in the North Sea area, major oil companies are already planning production from fields in depths of water in the 1000 to 2000 feet range in other parts of the world.

Obviously, drilling and production requirements will be very different from present techniques used for operating in shallower depths. It will be necessary to design and develop entirely new methods of operation and equipment to handle environmental, pressure and production problems which have never previously been experienced and which will demand an approach which, in our present situation, may appear to be bordering on the realm of science fiction.

Without going too deeply into the problems of the future we already have enough problems to tax existing expertise to its utmost if we are to drill, produce and pipe the known oil reserves in the North Sea area at a sufficient rate and financial consideration to satisfy the immediate needs of the United Kingdom, especially if we are to compete successfully with the importation of oil from overseas and the Middle East. It is a frightening thought to realise that overseas producers need only reduce the price of a barrel of oil by a few dollars, which they could easily accommodate, in order to create a situation where it would not be financially worthwhile to continue with North Sea operations.

In oil exploration the first step is for the geologist to identify an area where sedimentary rock formations of a suitable age exist and where, hopefully, reservoir conditions are suitable for the retention of oil and gas. The geophysicist then examines the identified area and attempts to identify anticlinal structures which are sealed by a cap rock and therefore present the most possible prospect of containing entrapped oil or gas. Unfortunately, neither the geologist nor the geophysicist is able at this stage to guarantee that hydrocarbons are actually present in a formation and consequently the only means of proving the prospect is by undertaking a very expensive drilling programme.

Even using the most modern equipment and technology, it is frequently necessary to drill some twelve to twenty wells in a wild-cat area before it is possible to evaluate the prospects of the existence or potential of a reservoir. Obviously, drilling costs are very much higher in regions of deep water than on land and a single well in a remote area or at sea may easily cost £1 million or more to drill.

The chances are that many dry holes will be drilled before a commercial find is established and consequently the petroleum engineer and the accountant are continually called upon the decide whether the expense of a drilling programme is justified and, if so, at what stage it should continue or should be abandoned if evidence does not support the hope of a discovery which will eventually provide a profit of reasonable proportions. This is unfortunately a situation which is not always fully appreciated by some governments who are inclined to examine the picture after a major discovery has been established, and then feel that they should be entitled to cash-in on the profits without accepting any of the risk costs. This approach has, in the past, been responsible for retarding progress in some overseas areas and in some cases has resulted in almost complete strangulation of the exploration programmes.

It is well to appreciate that major oil companies who are able to finance the huge projects involved are more than liable to invest their money in more profitable ventures in other areas if the authorities introduce conditions which are considered to be unreasonable and prevent a fair return on their investment. Like it or not, the major oil companies have access to finance and technical expertise which is almost unique and more than a few governments have, to their cost, found truth in the saying, 'You cannot fight a major oil company'.

The first oil well was drilled in August 1859 by 'Colonel' Drake in Titusville, Pennsylvania, using the cable tool or percussion method which made use of a drilling bit reciprocated at the end of a wire cable in order to pulverise the rock formation which was then bailed from the resulting hole by using a bailer or sand pump. The method was slow and dangerous as the open hole allowed gas or oil to flow freely to surface when the pay zone was penetrated and this situation often resulted in the gusher which frequently flooded the surrounding area or caused a major fire before the well could be brought under control.

Later refinements in the method gave greater control but it was not until the introduction of the rotary system of drilling and the use of mud control that wells could be completed in comparative safety and the accepted gusher became a thing of the past. Nowadays there is a saying that 'one never sees oil in an oilfield unless something has gone wrong!' This means, of course, that even very high pressure wells are drilled under control at all times and when production commences the oil or gas is piped to the tank farm or refinery without 'seeing the light of day'.

The rotary system of drilling began to take over from the cable tool method in the late 1920s and today it can be fairly said that all oil wells are drilled using the rotary system. The cable tool method is to this day often used in drilling water wells to shallow depths as although progress is slow the equipment is inexpensive and crews do not need to be so highly trained, nor is there normally any danger of a well flowing out of control with disastrous results.

A rotary drilling rig may be compared to the common drilling machine in a workshop which is used to drill holes in metal or other materials. Basic principles are the same but the problems encountered in drilling through the earth's crust to possibly 30,000 feet and to combat situations of unconsolidated formations in very high pressure shows of water, gas, or oil obviously presents problems which demand special equipment, materials and techniques not required under normal conditions.

The basic essentials for drilling any hole are a bit to attack the material being drilled, a shaft to connect the bit to the power unit providing rotary motion and to allow for a feed-off as the depth progresses; a means of removing the material cut by the bit, a power unit to supply the necessary torque and to remove the shaft and bit from the hole.

The bit used for drilling a known material can be easily programmed but, in the case of a well the type and hardness of the formation is continuously changing without the operator having any means of knowing what to expect or over what section the hole will follow a particular pattern. The complications presented by this situation can be imagined if it is realised that a bit which is suitable for attacking a soft formation is quite useless in a hard formation and a hard formation bit is equally useless in a soft formation.

Since it may take up to twelve hours or more to pull out and re-run the drill shaft to change a bit at say 8000 feet, and at a cost of some £15 per minute to operate a bit rig, it can be appreciated just how vital are the decisions which have to be made by the drilling superintendent, the geologist and the petroleum engineer.

The drill string, which serves to transmit rotary motion to the bit and provide passage for the circulating fluid, consists of 30 foot lengths of very special steel pipe fitted at each end with fast type tapered screwed connectors known as tool joints, which allow the pipes to be made up rather like a common chimney sweep's flue brush assembly to provide a drill shaft to whatever depth is required.

Immediately above the bit, heavy section pipes are run to provide weight on the bit and to hold the remainder of the drill pipe above in tension, so that a 'plumb bob' effect assists in keeping the hole following a vertical path. The top pipe of the drill string is some 35 feet in length and has an outside square or hexagonal section which provides a means of transmitting rotary motion to the drill shaft via the power unit and rotary table, and also allows the drill shaft to be raised or lowered when still or when being rotated.

The cuttings which are produced by the bit as it attacks the formation are removed to surface by fluid which is circulated through the drill pipe and returns in the annular space between the pipe and the well wall. This fluid which is circulated in the well is known as mud but is, in fact, a very specialised liquid, usually water or oil based, which is treated with a multitude of additives to accommodate conditions existing in the well being drilled. The mud also serves the purpose of cooling the bit, lining the well wall and consolidating unstable formations; reducing friction between the pipe and the well wall; retaining cuttings in suspension when circulation ceases for any reason; permitting cuttings to be deposited on return to surface and, by means of adjusting the specific gravity value, is capable of controlling any down hole pressure conditions due to the presence of fluid or gas in a reservoir formation.

The drilling string which, as mentioned above, is the pipe extending from the rotary table on the derrick floor to the bit attacking the formation, is suspended from a hook and pulley block system supported by the derrick structure and is capable of being raised or lowered and rotated by the main power unit operating a massive winch (known as the drawworks) and a rotary table also powered by the main engines.

The main equipment used for drilling a land-based well is essentially identical to that used in an offshore operation when drilling from a fixed platform but some special equipment is required when drilling with a semi-submersible unit or a drill ship, as provision must then be made to accommodate the motion of the vessel due to wave action.

When a semi-submersible or a drill ship is used the wellhead is set on the seabed, and compensating devices are required in the drilling string to ensure that the motion of the rise and fall of the vessel is not transmitted to the bit which is attacking the formation at the bottom of the well. To achieve this result the bit is loaded with a number of heavy section pipes known as drill collars to give the required bit loading. The drill collar assembly is attached to the main drill pipe string by means of expanding or telescopic pipes, which allow the surface unit to rise or fall over a predetermined height without causing the drill collar assembly and the attached rock bit to be lifted from or pressed into the rock formation.

Additional compensating equipment is also needed to maintain a steady tension on the riser pipe, which extends from the wellhead assembly on the seabed to the underside of the surface unit, and which serves to handle the mud returns from the well to the mud screens located on the floating rig.

Although the hydrostatic head of a circulating mud column in the well is the first line of defence against any fluid pressures which may be present in down hole reservoir formations, it is essential to provide packer assemblies at the wellhead which are capable of closing in the well either when the drill pipe is in the hole or when the pipe is out of the hole. The assembly so fitted is known as a blow out preventer (BOP) and the closing elements are normally controlled by hydraulic pressure.

Because of their vital importance all blow out preventer stacks are precision made and tested under the most stringent conditions to withstand any pressure ratings which can possibly be foreseen. Where blow out preventer stacks are installed on the seabed they must be equally reliable, in spite of their inaccessibility from a maintenance, operational and inspection point of view. Such units may be some forty feet in height and incorporate a series of ram elements which are capable of closing in on varying diameters of drill pipe or casing or on an open hole. Special cutting rams are usually fitted which can shear through a drilling string if a situation should arise where it is necessary to drop the drill pipe into the hole for the purpose of closing in a dangerous well.

Such sub-sea BOP installations are capable of being installed, tested and operated from a control point on the derrick floor, and an indicator panel at the drillers position provides information on the wellhead situation at all times. Frequently a floating rig is also provided with a control panel on the bridge to allow the 'Master of the Ship' to close in the well and abandon the site, if he considers that sea conditions or operational conditions present a serious hazard to the vessel under his command.

Possibly the item of equipment which presents the major problem when using a floating rig is the riser assembly, which is a large diameter pipe (16 inch to 20 inch in diameter), attached to the wellhead on the sea floor by a hydraulic connector and extending to a few feet below the derrick floor of the floating rig. The riser is assembled in sections and serves the purpose of providing a passage for the mud returns from the well to return to the vibrating screen, desanders and desilters (which remove formation particles), before the circulating fluid returns to the well via the circulating pumps and the drill string. The riser assembly is provided with a telescopic section to accommodate the rise and fall of the vessel and universal couplings which permit some drift from the true vertical over the wellhead but to a maximum angle of 10 degrees.

Running or recovering the riser assembly is a very time-consuming operation and many hours of rig time can be lost if a situation arises whereby it is necessary to raise the blow out preventer stack from the sea floor for repairs or maintenance or for any other reason. In high storm conditions the rig anchors may drag and the riser assembly must be disconnected from the wellhead if its maximum angle of 10 degrees off vertical is exceeded, otherwise the wellhead assembly would be subjected to dangerous stress.

Added complications are involved in running or recovering the riser assembly as two high pressure pipes of two inches or so in diameter must be made up or recovered at the same time. These lines are known as the kill line and the flow line. The kill line is connected to the wellhead below the blow out preventers and is used for pumping mud to kill a well which is closed in and under pressure. The flow line provides a means of producing fluid or gas from the well when the main wellhead valve is closed. These two pipelines are run in lengths of around 30 feet in conjunction with the riser assembly and they must withstand a pressure test of at least twice the estimated pressure which a well may generate with all wellhead valves closed.

Once the floater or semi-submersible has completed the exploration programme by drilling and testing a series of wells and a field has been established, it becomes necessary to install a production platform from which as many as 24 to 30 wells may be drilled at angles to provide a means of producing the reservoir fluids from possibly a two mile diameter area.

Production platforms are very expensive indeed to assemble and to position on site. A single platform under North Sea conditions, with necessary production equipment installed, may cost in the region of £60 million at current prices. Platforms come in two main types, steel structures or cement structures, and each has its own sphere of usefulness according to the seabed conditions. Steel platforms are fixed in position by piling often to a depth of 250 to 300 feet; this alone is a major operation, to say nothing of towing and positioning under weather and sea conditions which are likely to be anything but ideal.

Never before, anywhere in the world, have production platforms been installed in depths of water and weather conditions such as are experienced in the North Sea. This situation must be a cause for concern, as the designer and the engineer have no means of testing the completed structure under operational conditions before its installation. One is tempted to compare the exercise with the idea of constructing a new design of aircraft and putting it into operation without previously subjecting it to an extensive test programme.

Within the next decade there will be some eighty production platforms in the turbulent conditions of the North Sea each of which will be the focal point to serve some twenty-four to thirty-six wells on full production. Hopefully, no platform will ever suffer collision by a ship; a structural failure; enemy action; sabotage or even failure due to unforeseen storm conditions. In theory, a disaster of this nature would not result in any serious pollution problem, as each well is fitted with shut off devices to close in the production in the event of damage to the wellhead. Unfortunately, safety devices can fail and the best of theories can prove wrong.

In this connection Mr John Smith, Under-Secretary at the Department of Energy, in a written reply in the Commons stated that an offshore blow out would require ninety days to bring under control. Many oil men would consider this opinion as optimistic but, in any case, the thought of a wild well flowing oil into the sea for any length of time hardly bears thinking about.

At the present stage of technology there is little alternative to the use of fixed platforms and directionally drilled wells if the oil is to be brought ashore quickly, but from a safety point of view it would be much more satisfactory if vertical wells could be drilled in a pattern remote from the production unit and completed with the wellheads on the seabed, and piped through a sub-sea collecting station to the production platform so that any disaster situation involving the platform would not result in damage to the wellhead. This type of completion is already available for medium water depths but it may be some time before the system is adaptable to water depths of below 300 feet.

A well blow out on land can present very serious problems, often involving the drilling of a directional hole from some remote position to intersect the wild hole for the purpose of pumping mud or cement slurry to bring it under control. To kill a wild well at sea in this manner would be a much more difficult exercise, as the intersecting hole would be aimed at a directionally drilled well almost impossible to locate with any degree of accuracy.

There is also the possibility of a major fire on one of the production platforms; a highly complex problem particularly at a time when sea and weather conditions may make access to the site difficult, if not impossible. No doubt much expert thought is being given to these and like situations, but it would seem that other and safer methods of bringing a field on to production should be aimed at.

The accepted view in the industry is that technological and economic factors will prevent the installation of production platforms in water depths of more than around 600 feet. Oil operators are, however, already looking at the prospects of exploring for reservoirs in much greater water depths and there is no doubt that alternative production systems must be designed to accommodate any successful finds at these depths.

Many other problems are involved in offshore drilling and production operations. The supply of equipment, materials and food to an exploration drilling rig is a major exercise when it is realised that one hole may require 500 tons of casing and tubulars; 900 tons of bulk chemicals; 250 tons of bagged cement; 900 tons of fuel; 150 tons of food and general stores; 200 tons of tools and contingencies. Relate this scale of figures to a production drilling platform with two rigs in operation and 27 to 30 wells to be drilled, and the necessity for maintaining an uninterrupted operation, and one can appreciate the difficulties facing the supply organizations, particularly as weather conditions so often disrupt the supply lines for days or weeks at a time.

Similar problems face the engineers when the production equipment has to be transported, offloaded on to a platform and installed after the production wells have been completed. Some of the module loads to be handled are in the region of 2000 tons weight and delays due to weather conditions can result in months of delay in bringing the oil ashore.

The only reliable way to bring the oil ashore on a continuous programme from a distant offshore field is to install a large diameter (32 inch to 36 inch) pipeline. Pipelines of this size and length have never previously been laid, and here again is a new field of engineering requiring very special and expensive equipment and techniques which are still in the experimental stages.

Reservoir oil almost always contains large volumes of gas which expands as the pressure falls when the fluid flows to surface. The gas must be separated at the wellhead before the oil is pumped ashore through the pipeline and herein lies another problem facing the production engineer. If the gas/oil ratio is small it may be possible usefully to employ the gas for running the power units on the production platform and to dispose of any small surplus by flaring. When the gas/oil ratio is high, however, there may be many millions of cubic feet of surplus gas available per day and flaring cannot be considered if only on account of the tremendous waste of valuable energy. If there is sufficient gas to justify the laying of a pipeline to shore and if a ready market for the gas exists, then this may be the best solution.

In many cases, however, one or other of the above conditions may not exist and in that situation the surplus gas may be re-injected into the down hole reservoir via wells specially drilled into the gas cap. Compressors for handling very large volumes at high pressures are required to undertake this operation, but the re-injected gas helps to maintain the field production pressure and the wasteful practice of flaring is avoided.

Although the basic methods used for drilling an onshore well are the same as those used for an offshore well, it will be appreciated that the factors introduced by increasing water depths, weather conditions, supply problems and the handling of any fluid discoveries greatly complicate the entire exercise and so increase the operational costs that only potential areas for large production become of interest to the exploration companies.

Since the general adoption of the 'rotary system' of well drilling in the 1930s, the basic method has changed very little over the years. Diesel power and diesel electric has replaced steam as a source of power on the drilling rigs and obviously there have been big advances in steel and more sophisticated materials which are available for the manufacture of equipment. Probably the most spectacular advances have been in mud control for drilling and in electric logging of wells to provide information on the types of formation penetrated and their potential as oil or gas reservoirs. However, there are still tremendous opportunities available for new ideas and for the design of equipment in an industry which already is reaching beyond the technically possible.

OIL TERMS

A

abandon: to cease work on a well which is non productive (a dry hole) and to plug off the well with cement plugs *(see cement plugs)* and salvage all recoverable equipment.

absorption: the accumulation of a thin layer of molecules of gas or liquid in a solid surface.

accelerator: cement slurry is used for cementing casings in a well or for other purposes such as sealing loss circulation zones *(see lose returns)* or placing cement plugs *(see cement plugs)* on completion of a 'dry hole' *(see duster)*. In order to reduce the setting time of the slurry, when this is advantageous, an accelerator is added to the slurry. Chemicals which act as accelerators are: calcium chloride, sodium chloride.

accumulator: an accumulator on an oil rig *(see rig – drilling)* is used to actuate the blow out preventer fittings on a well head *(see blow out preventer stack)*. In effect the accumulator is a pressure container charged with nitrogen

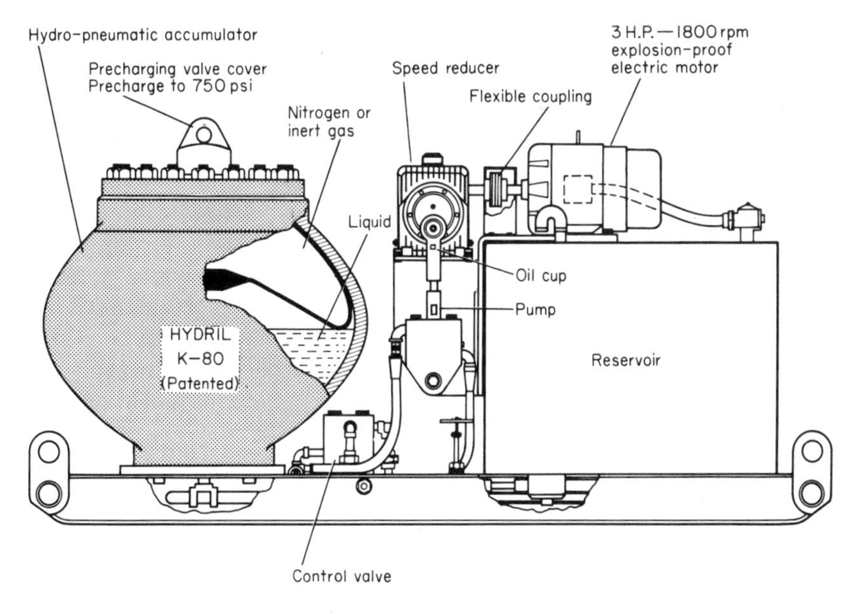

Electric Motor Driven Accumulator Unit

and into which fluid is pumped, thus compressing the nitrogen to a high pressure (1500 to 2000 p.s.i. working pressure). When the control cock to the operating manifold is opened the hydrogen forces the fluid out of the accumulator and operates the well head packers

acid bottle: a glass tube which is run into a hole in a steel container to record deviation from the vertical by the use of hydroflouric acid which attacks the glass and etches a line from which the angle of deviation can be ascertained. The acid bottle is still used in mineral drilling operations but has been replaced in oil well drilling operations by much more sophisticated instruments such as 'Totco' recorders or the Eastman Camera.

acidize and **acid treatment:** the technique of increasing production from a limestone reservoir by pumping acid into the formation to increase the permeability *(see permeability)* in the region of the well bore. Limestone reservoirs may lack good permeability properties and so may resist the flow of oil into a well bore. Production volumes can often be greatly increased by pumping acid into the formation to attack the limestone and provide easier passage for the oil to flow into the well bore.

air lift: the technique of injecting compressed air into the fluid column in a well to stimulate fluid flow *(see gas lift).*

anchor: when a well penetrates a formation which appears to contain oil or gas it is frequently decided to test the potential production rate of fluid from the reservoir. The method used is to set a packer *(see packer)* above the formation to be tested and, by so doing, to support the mud column above the packer and thus to reduce the hydrostatic pressure on the reservoir to permit any fluids present to flow into the well bore. An anchor pipe (perforated over a section) is run below the rubber packer element and this pipe extends to the bottom of the well and provides a means of expanding the rubber element by lowering the weight of the drill pipe above the packer. Once the packer element is expanded to seal against the well wall or casing, fluid from the reservoir formation can flow through the tail pipe perforations onto the drill string and so to surface for analysis and flow rate estimating.

anchor (dead line): the fitting which anchors the 'dead line' of the hoisting cable which accommodates the loads handled by the drawworks *(see drawworks).* This anchor is usually fitted with a diaphragm device which indicates on a gauge in front of the driller the loads which are being handled by the hoisting equipment.

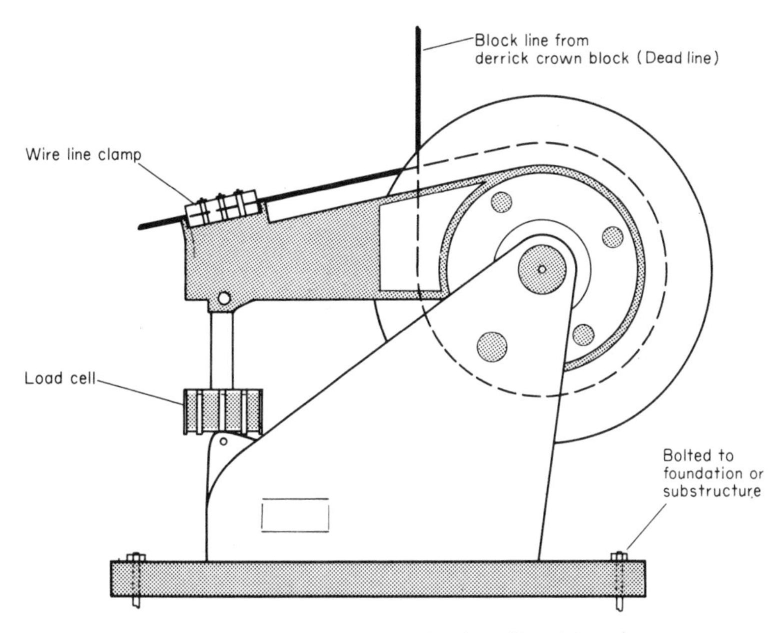

Dead Line Anchor — or Anchor (Dead Line)

angle of deviation: the deviation from true vertical which a bore hole may suffer either by accident or by design.

annular space: the space between the drill string and the well wall, or the casing string and the well wall.

associated gas: natural gas in a reservoir formation which overlies and is in contact with the crude oil.

B

back off: usually refers to the unscrewing of drill pipe from a 'fish' *(see fish)* in the hole.

back pressure: pressure resulting from restricting the full natural flow of oil or gas.

back up man: member of the drilling crew who holds the tong to prevent a length of pipe rotating whilst another length is screwed on or out of it.

bails: links which connect the main hoisting hook with the drill pipe elevators.

bailer: a cylindrical container fitted with a foot valve which is used to remove fluid or slurry from a hole and is run on a wire line. A bailer is used in a cable tool drilling operation *(see drilling – cable tool)* to remove slurry and cuttings produced by the bit as it attacks the formation. A bailer may also be used in a

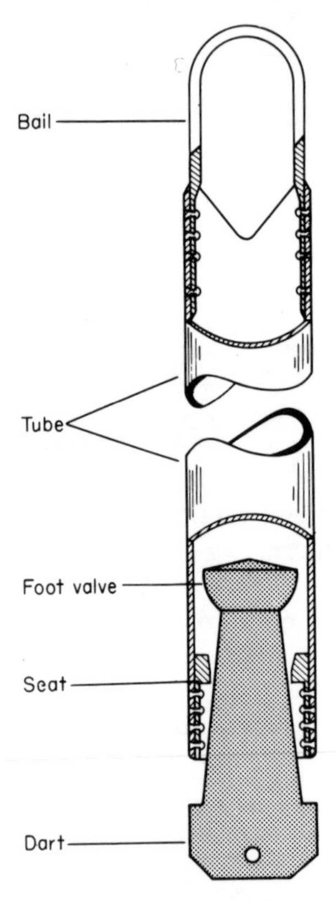

Bailer

well drilled by the rotary method *(see drilling – rotary)* in order to lower the level of the mud column to reduce the hydrostatic pressure on the reservoir formation to a degree where fluid will flow into the well bore.

ball up: the situation where 'sticky' formation such as 'gumbo shale' chokes the bit cutters and makes a round trip *(see round trip)* necessary to clean the bit.

ball weevil: an inexperienced oil field worker.

bare foot completion: the completion of a well in a reservoir formation which is stable and does not require a liner or perforated casing completion *(see completion).*

barite: barium sulphate with a specific gravity of 4.2 which may be added to the drilling mud in circulation to increase its weight and provide a hydrostatic head for controlling formation pressures.

barrel: one barrel of oil equals 35 Imperial gallons or 42 U.S. gallons. The barrel is the generally accepted measurement when describing the production potential of a well or when measuring mud volumes.

barrel (core): a tubular device fitted at its extremity with an annular type bit *(see core bit)* and designed to recover a solid bar of the formation being drilled. Such cores inform the geologist in much more detail of the formation being drilled than is provided by cuttings recovered from the mud in circulation as it passes over the vibrating screen *(see vibrating screen).* Core barrels may be 30 ft to 60 ft (9.1 m to 18.3 m) long and usually have an inner and an outer barrel or tube. The core bit is screwed to the outer barrel which is rotated by the drill string as in a normal drilling operation. The inner barrel is mounted on bearings and does not rotate as the core enters. The diameter of the core being cut depends upon the size of the central hole in the bit. When the coring operation has been completed the barrel is pulled up and the core breaks from the formation at the bottom of the hole and is retained in the inner core barrel by the core catcher or retaining ring fitted at the lower end of the inner barrel.

barrel wrench: a special 'friction wrench' used for repairing a down hole pump *(see down hole pump).* The barrel of such a pump is very thin and must be handled with extreme care in order to avoid crushing or similar damage.

basket: a fishing device *(see fishing tools)* for recovering 'junk' from the bottom of a hole such as roller bearings or bit cutters from a collapsed rock bit. The cylindrical tube will usually be fitted at its lower end with a cutting head above which is a ring fitted with spring loaded fingers which serve to retain any 'fish' recovered.

basket (cement): a cement basket is a funnel shaped rubber bucket which is provided with spring type fingers to hold the upper rim against the well wall. A basket is run on a string of casing *(see casing string)* when it is desirable, for reasons such as protecting a formation from contamination, to cement the casing only above the point at which the basket is set. The cement slurry is pumped down the inside of the casing string and passes out into the annular

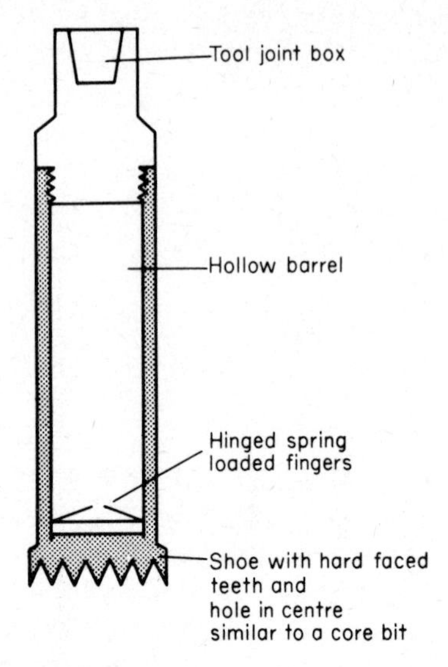

Junk Basket

space between the casing and the wall of the well through suitably placed holes above the cement basket.

bean: the name 'bean' is given to a choke device which is used to control the flow of fluid or gas under pressure through a pipe line. An adjustable bean has a hardened steel replaceable seat tube with a tapered needle controlled by a hand wheel in a similar manner to a valve.

Other 'beans' are hardened steel tubes with an accurately machined bore which are inserted into a pipe system to control the fluid flow.

bentonite: a colloidal clay composed of montmorillonite and which swells when wet. Bentonite (silicate of Ca, Mg and AC with H_2O) is the essential basis of most drilling muds and imparts properties to the fluid such as gel forming thixotropic, non-corrosive, non-abrasive and lubricating. A bentonite mud also has good filtration properties which is important when drilling in porous rocks as there is no serious build up of filter cake on the well wall due to water loss to the formation.

bird cage: to flatten and spread the strands in a wire rope.

bit breaker: a plate which fits into the master bushing recess in a rotary table and enables drilling bits to be unscrewed from the drill string.

bit collar: heavy duty pipe which is used to connect a rock bit to the drill string.

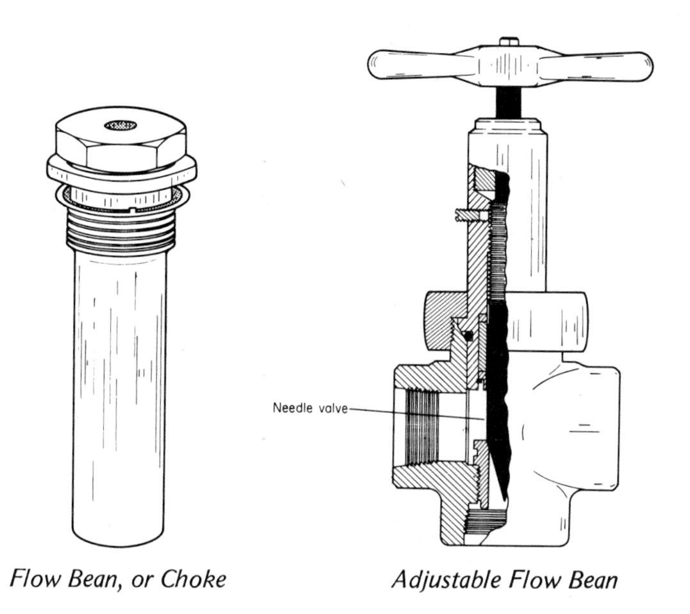

Flow Bean, or Choke   *Adjustable Flow Bean*

blank liner: a liner *(see liner)* which has no perforations.

blind rams: steel rams with rubber inserts which are fitted to a blow out preventer *(see blow out preventer stack)* and which, in the closed position, will shut in the well head in a similar manner as a main valve.

block line: the wire line which is spooled on the main drum of the drawworks *(see drawworks)* and is reeved over the crown block *(see crown block)* sheaves and through the travelling block *(see travelling block)* for the purpose of handling the drill string or other loads. The wire diameter is usually 1⅛ inch or 1¼ inch (29 mm or 32 mm) and may have a wire core.

blow out: a situation where a well becomes out of control due to the fluids from the formation 'blowing wild' at the surface. The cause of a blow out may be one of many, such as sabotage, failure of well head equipment or down hole equipment, carelessness on the part of the crew or an 'act of God'. Fortunately 'blow outs' are rare but when they do occur they can be disastrous.

blow out preventer stack: an assembly of control gates fitted at the casing head which are capable of closing around drill pipe or casing and are designed to control a potential 'wild well'. The components of a blow out assembly are

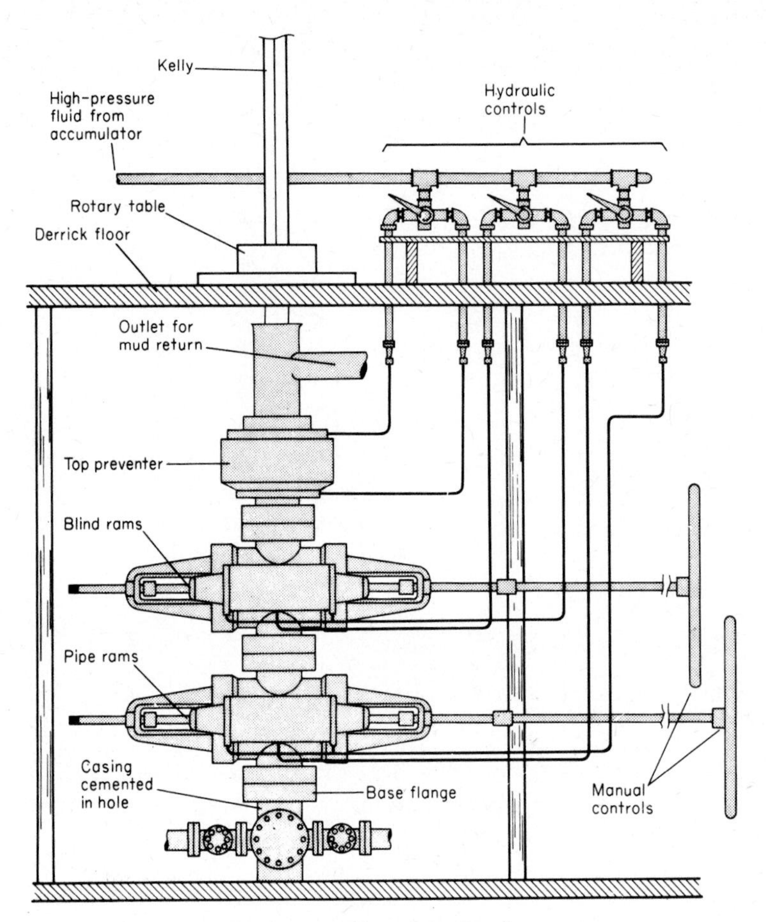

Blow out Preventer Stack

manufactured to the highest possible standards and may in some cases be tested to 20 000 lbs/in^2 (1 406 kg/cm^2).

board (fourble): a board or platform fixed in the derrick at a height of some 90 feet (27.4 m) above the floor to facilitate the handling of drill pipe stands being run in or pulled out of a well.

board (head): protection board over the head of the drillers' position.

boilerhouse: to make up or fake a report without actually doing the work.

borehole: a well *(see well).*

bottom hole differential pressure: *(see differential pressure – bottom hole).* In some wells such as those in the big fields of the Middle East, bottom hole differential pressures may amount to 4 000 lbs/in^2 to 5 000 lbs/in^2 (281 kg/cm^2 to 351 kg/cm^2) or more but in many areas the pressure may be negligible.

bottom hole pressure: the pressure existing at the bottom of a hole.

bottom water: water occurring in the formation below the oil in a producing sand or from a sand below the producing sand.

bowl: a heavy steel ring into which fit tapered slips *(see slips)* to support a tubing string.

bowl (casing): a device for repairing a damaged casing string.

bradenhead gas: or 'casing head gas' i.e. gas produced with oil or from the casing head of an oil well.

breaking down: the operation of unscrewing a drill string into 'singles' *(see single)* when a well has been completed and the drill string is pulled out for the last time.

breakout: the operation of unscrewing joints of pipe from the drill string.

bridge: an obstruction in the hole due to caving formation or some similar cause such as the presence of a 'fish' *(see fish).* A bridge may be removed by cleaning out the hole with a rock bit and reamer assembly *(see reamer)* or some cases may demand the use of a milling tool or other fishing device according to the cause of the obstruction.

bridge plug: a packer assembly fitted with slips and a rubber sealing sleeve which is run into a hole to isolate a down hole producing zone in order to test a zone at a higher level.

bridging material: fibrous material which is added to the 'mud' *(see mud)* to seal a 'loss circulation' formation *(see lose returns).* All sorts of materials are used for this purpose such as cotton seeds, chopped hay, chopped palm leaves, sawdust, straw, cellophane strips, torn up rags or sacking – in fact anything which may help to seal the thief formation

bring in: the process of causing fluid to flow into the well from the formation and thus 'produce the well'. The aim of the exercise is to reduce the hydrostatic pressure at the reservoir face, due to the fluid column in the well bore, to a degree where the formation fluid will flow into the bore. The specific gravity of the mud may be reduced, or the mud column may be replaced by water, in order to reduce the hydrostatic head. Other methods used are to bail fluid from the well using a bailer *(see bailer)* or to swab the hole using a swab *(see swab).*

bull rope: an endless rope used in cable tool drilling *(see drilling – cable tool)* to drive the 'bull wheel' *(see bull wheel).*

bull wheel: winding drum assembly used in cable tool drilling *(see drilling – cable tool).*

bushing: steel inserts fitted into a rotary table *(see rotary table)* to accommodate the kelly drive bushing *(see kelly bushing).*

bushing (kelly): *(see kelly bushing).*

bushing master: *(see master bushings).*

C

cable tool bit: chisel type bit used in cable tool drilling *(see drilling – cable tool).*

cable tool drilling: *(see drilling – cable tool).*

cake (mud): *(see mud – cake).*

calf line: wire line used for lowering casing into the hole when cable tool drilling *(see drilling – cable tool).*

calf wheel: hoisting drum on a 'cable tool rig' *(see drilling – cable tool)* which handles the 'calf line' *(see calf line).*

calliper log: an electric log which records on film the variations of diameter of a bore hole from total depth to surface. In hard formations the drilling bit will cut a circular hole of true diameter but when attacking softer formations or salt deposits the diameter of the well may be greatly enlarged due to caving of the walls or leaching out of a salt deposit by the mud in circulation.

In order accurately to estimate the volume of cement slurry needed to cement a casing string *(see casing string)* it is important to know the extent of such enlargements in the hole diameter and this information can be obtained by running a 'calliper log'. A calliper tool, having spring loaded arms which contact the well wall, is run into the hole on an electric cable and the contour of the walls of the bore is recorded on a graph as the tool is run from total depth to surface. From this diagram it is possible to estimate the true volume of the well bore.

cap rock: an impermeable layer of rock which overlays an oil or gas reservoir and prevents migration of the fluids.

casing: steel pipe used to line a well after drilling the hole.

casing (land): *(see land – casing).*

casing bowl: *(see bowl – casing).*

casing clamp: a clamp which fits around the casing being run in or pulled out of a hole.

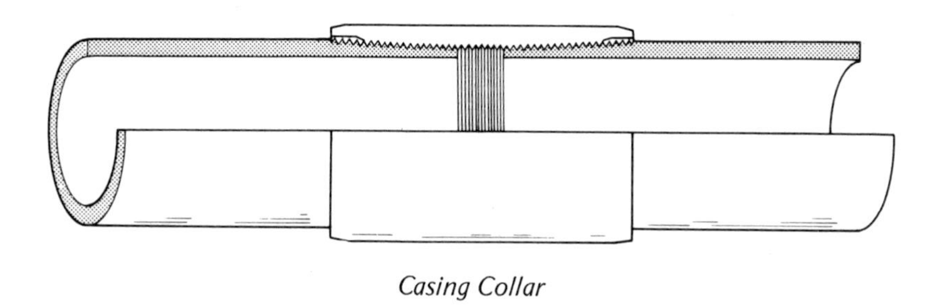

Casing Collar

casing collar: a collar screwed to a casing joint for connecting the next joint to be run.

casing cutter: a device which is run on a drill string and has hardened cutters controlled by fluid pressure which are used for milling through a casing string. A casing cutter may be used as a fishing tool *(see fishing tools)* in order, for example, to cut and recover casing in lengths from a casing string *(see casing string)* which is stuck in a hole.

casing line: steel line used on a cable tool rig *(see drilling – cable tool)* for running a casing string *(see casing string)*, otherwise known as a 'calf line'.

casing perforator: device for making perforations in a casing string *(see casing string)* opposite an oil zone to allow production to flow into the casing.

casing pressure: gas pressure built up between a casing string *(see casing string)* and a tubing string.

casing protector: rubber sleeve fitted to the drill string to reduce wear in the casing and the drill pipe joints.

casing pump: down hole pump fitted in the casing to pump a non-flowing well. Many wells do not have sufficient reservoir pressure to cause the oil to flow to surface under its influence alone and it is, therefore, necessary in such cases, to install a down hole pump. Such a pump is run and set in the casing in the oil column and may be powered by an electric motor or more usually by sucker rods *(see sucker rods)* which are reciprocated by a pumping jack *(see pumping jack)* installed over the well head. The power requirement for the pump jack is small (10 H.P. to 20 H.P.) according to the depth of the hole and weight of the sucker rod string to be handled. The pump jack drive may be electric motor, diesel engine, petrol engine, or by gas recovered from the oil being pumped. Once installed these units require very little attention and they can be seen 'nodding away' in isolation in fields, jungles and deserts, pumping a few gallons at each stroke and only requiring a visit from the service engineer about once a week.

casing rack: rack usually made of steel pipes and located outside the derrick floor on which casing lengths are stacked before running into the well.

casing ramp: steel or wooden ramp from the casing rack to the derrick floor to facilitate pulling casing lengths into the derrick.

casing shoe: heavy section steel tube fitted to the lower end of a casing string *(see casing string)* to protect the end of the string from damage when running into a well.

casing spear: a 'fishing tool' *(see fishing tools)* which can be set inside a strong of casing *(see casing string)* to recover a dropped string or similar situation.

casing string: the term used for the steel tube which lines a well, after it has been drilled, and is made up of sections of pipe 20 ft to 30 ft (6.1 m to 9.1 m) in length and screwed together. A deep well , 15 000 ft to 30 000 ft (4 572 m to 9 144 m), may have as many as five strings of casing cemented in situation.

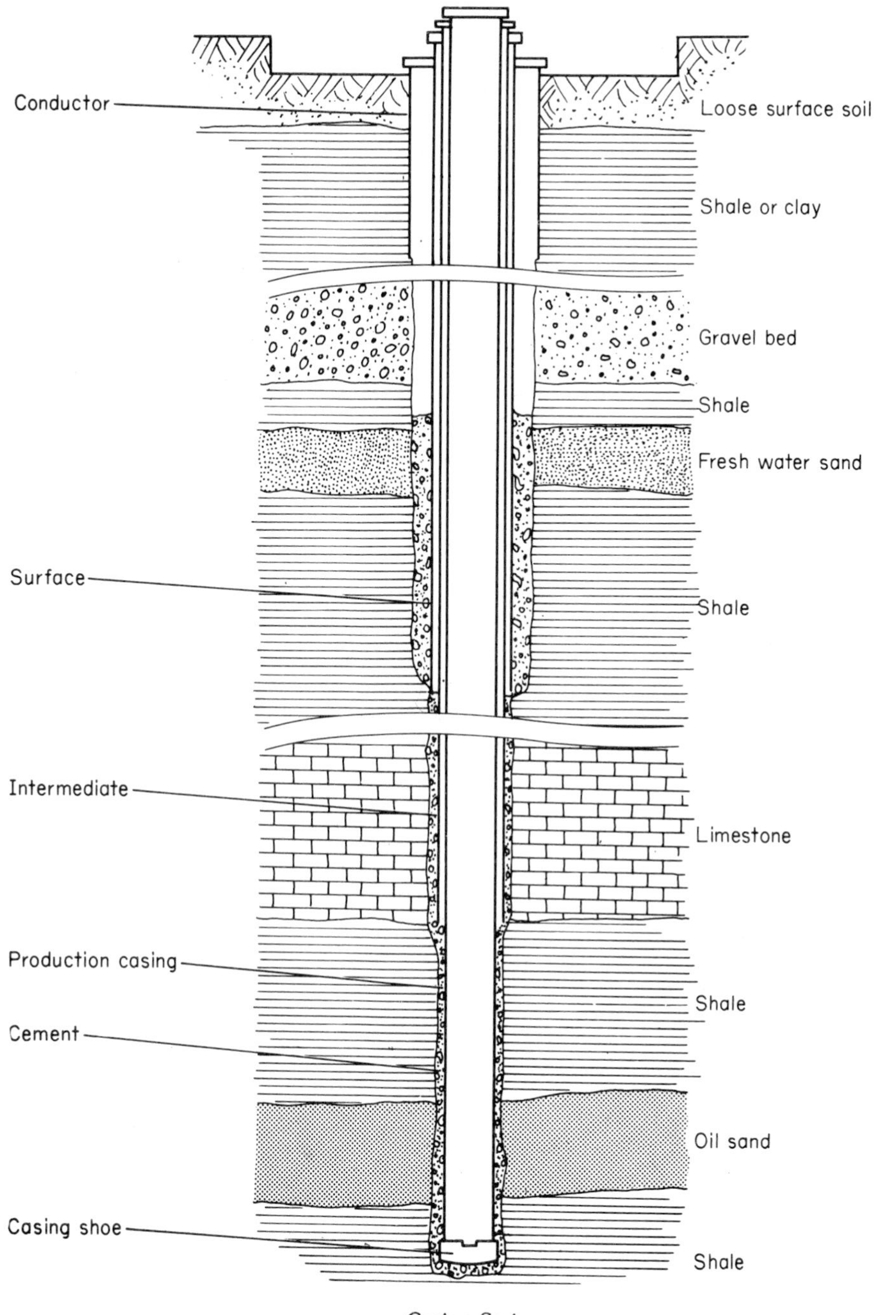

Casing String

A casing programme is designed before drilling commences and the programme is based on geological information, information from wells previously drilled in the area, if any, and an assessment of hole conditions which may occur during drilling but which are unknown until the situation arises.

Designing a casing programme is one of the most important aspects affecting the drilling and completion of a well and demands the expertise of the geologist, the drilling engineer and, in fact, all departments likely to be involved in the siting, drilling and production of the well.

Obviously there is no standard casing programme, but a typical example for a deep well, 15 000 ft to 20 000 ft (4 572 m to 6096 m), may be:

30 inches casing to 100 feet
20 inches casing to 500 feet
13³/₈ inches casing to 8 000 feet to 10 000 feet
9⁵/₈ inches casing to 15 000 feet
7 inches casing to final depth.

casing tester: a packing device used to locate leaks in a casing string *(see casing string)*. A packer assembly *(see packer)* is set in the casing at a chosen point indicated by the suspected or known failure and fluid pressure can then be applied to ascertain if, and where, a leak occurs.

casing tong: heavy duty adjustable type of wrench which hangs in the derrick and is used to tighten casing joints when running a casing string *(see casing string)*.

cathead: a bollard on the drawworks *(see drawworks)* cathead shaft used for handling a rope to pull pipe or casing tongs when making up or breaking out lengths of drill pipe or casing.

cat line: a rope or combination of a rope and wire line operated by the cathead and used for lifting equipment on the derrick floor.

catwalk: a ramp connecting the rig floor to the casing rack which provides a means of pulling drill pipe or casing joints on to the rig floor.

caving: situation where formation from the well wall caves into a hole.

cavity: enlargement of the hole due to caving or wash out of a soft formation. In extreme cases the hole may be plugged back *(see plug back)* with a cement plug *(see cement plugs)* and re-drilled. The possibility of such conditions in a hole makes it desirable to run a calliper log *(see calliper log)* before cementing a casing string *(see casing string)* in order to arrive at an accurate assessment of the volume of slurry required.

cellar: the excavation made before the drilling of a well which provides space for the installation of the surface well head equipment. The depth of the cellar will depend upon the height of the derrick substructure in use and the drilling well head assembly which will be required to handle the casing programme *(see casing string)*. Normally a cellar will be some 5 ft to 6 ft (1.5 m to 1.8 m) in depth.

cement head: removable head fitted to the landing joint *(see landing joint)* of a casing string to facilitate a cement job *(see cement job)*.

cement hopper: *(see hopper – cement)*.

cement job: the operation of cementing a casing string *(see casing string)* in the hole or setting cement plugs *(see cement plugs)*.

cement plugs: the term cement plug is used for a column of cement which is placed in a well bore for one of many reasons, and a plug may be a few feet in length or may extend over some hundreds of feet. Plugs may be placed to seal off porous formation zones which allow the circulating mud to flow into the formation. Or it may be necessary to 'plug back' *(see plug back)* a hole which has deviated from vertical in order to re-drill and correct the angle of deviation.

On completion of a 'dry hole' *(see dry hole)* cement plugs are run to isolate porous formations which could allow contamination of fresh water reservoirs due to the intrusion of salt water or possibly small gas or oil shows from formations which have been penetrated during drilling. It is common practice to set a plug at the bottom of a 'dry hole', one at the shoe of the last casing string and one from surface to around 50 ft (15.2 m) in the top casing string.

cement squeeze: forcing cement slurry into a formation with high pressure pumps. The term 'squeeze job' applies when cement is pumped under pressure in excess of that required to displace the slurry from the pipe or casing transmitting it into the well. Squeeze jobs provide a means of sealing formations which may contain gas, oil or water or for other purposes such as supporting an unconsolidated rock or for repairing an unsatisfactory casing cement job *(see cement job)*. The pressure involved is dictated by the bursting strength of the tubing or casing set in the hole and may be up to 5 000 lbs/in^2 (352 kg/cm^2).

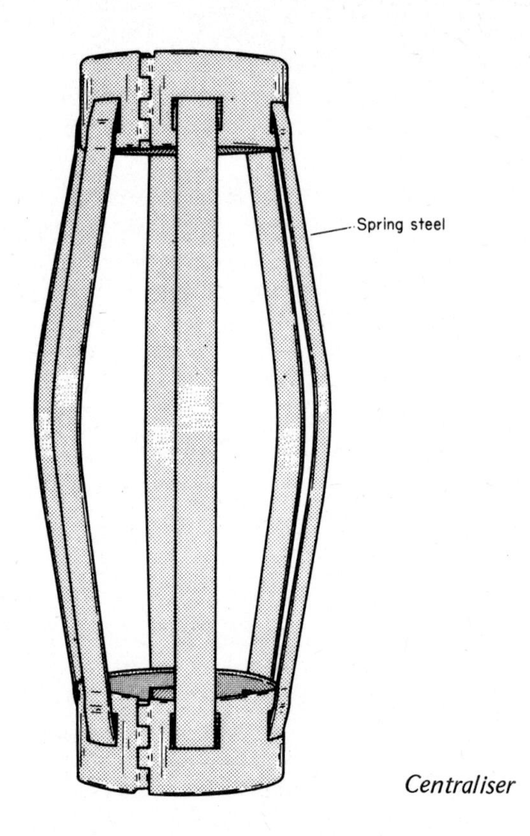

Centraliser

centraliser: a fitting which is placed on a length of casing to hold it centrally in the hole in order to ensure a uniform sleeve of cement around the casing joint.

channelling: a condition where cement around a casing string *(see casing string)* is not uniform and is channelled allowing fluid or gas to escape towards the surface. This condition may, in extreme cases, be remedied by squeezing cement into the annular space under high pressure.

cheater: a length of pipe which is slipped on to the handle of a spanner or wrench to give extra leverage.

chert: a rock formation which is harder than flint. Chert is a rock composed of non-crystalline silica, hard and tough and is the most difficult rock to drill.

chert clause: a clause in a drilling contract which stipulates that normal contract rates do not apply when chert formation is encountered.

choke: a removable steel orifice device fitted to a well flow line *(see flow line)* to restrict the fluid flow.

Christmas tree: the assembly of fittings and valves on a final casing to control the production rate of oil production.

circulating fluid: *(see mud).*

circulating head: a swivelling attachment which is crewed to a string of drill pipe or casing to permit pumping circulating mud or fluid into the pipe whilst at the same time allowing the pipe to be rotated and raised and lowered.

clean out: repair or cleaning operation in a well bore. The need for a 'clean out' job may arise due to deterioration of the formation or by 'waxing up' of the tubing in a production well which produces oil having a high wax content.

close in: a well which is capable of producing oil or gas but which is temporarily closed in at the well head.

closed in pressure: the pressure at the well head when all production valves are closed. The pressure will vary widely from field to field and may be 'zero' in some areas, and up to 5 000 lbs/in^2 (352 kg/cm^2) or more in areas such as the Middle East.

collar (bit): heavy duty pipe which is used to connect a rock bit to the drill string.

collar (drill: *(see drill collar).*

collar (float): *(see float collar).*

come in: situation where fluid or gas enters the well from the formation.

come out of the hole: pulling the drill string out of the hole.

commercial production: production capacity which will show a financial reward. Obviously a well drilled in an expensive area such as at sea or in a remote desert or jungle area must product a much greater volume of oil to be commercially productive than is the case close to a populated area. In the latter case a few barrels per day (say 100) may be worth while whereas offshore wells in the North Sea or wells in the Middle East would be required to produce several thousands of barrels per day to be considered as commercially productive. Many Middle East wells will produce 40 000 barrels per day for many years. The value of commercial production is not judged on the results of one well but on the capacity of a field which is estimated from results from several wells.

COMPLETION

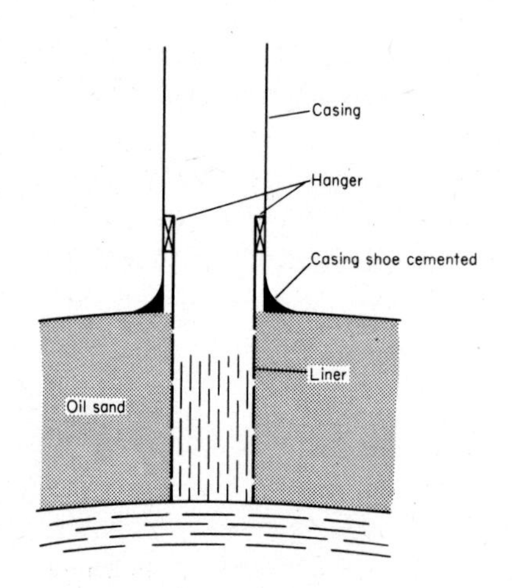

Slotted Liner Completion

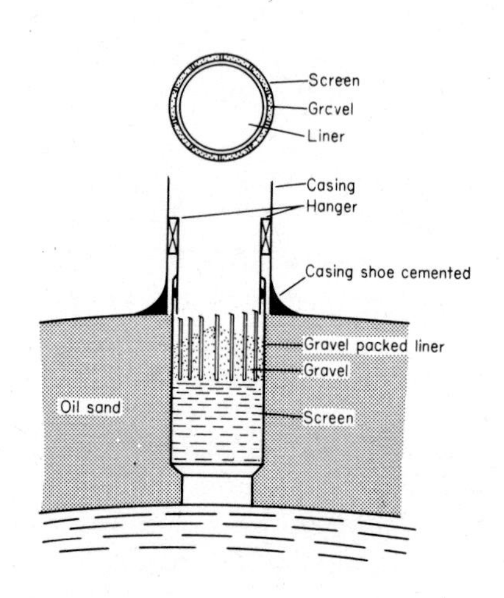

Gravel Packed Completion

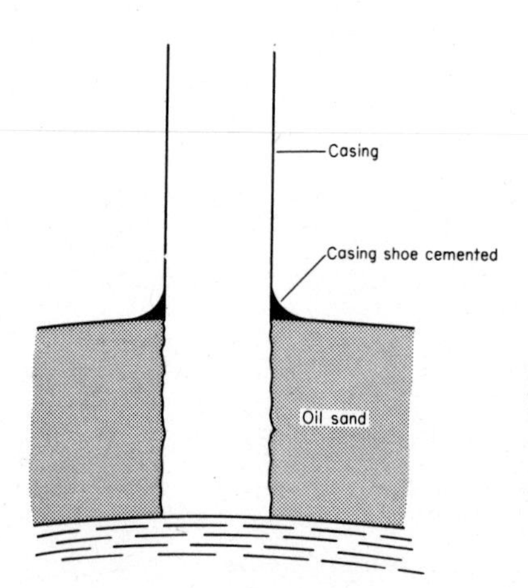

Bare Foot Completion

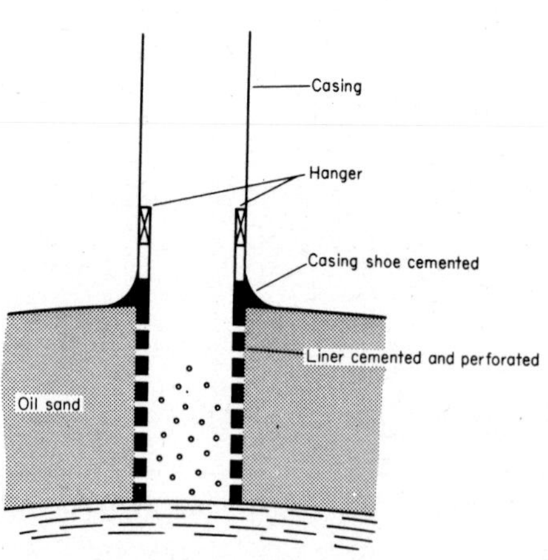

Liner Completion

completion: this term refers to the installation of permanent production equipment for the production of oil or gas.

condensate: hydrocarbons which are in the gaseous state under reservoir conditions and which become liquid either in passage up the hole, or at surface due to the reduced pressure conditions.

conductor: large diameter pipe which extends from underneath the rotary table to the starting point of the well and which provides a means of returning the circulating fluid to the mud screen on its return from the well bore.

cone bit: type of rock bit having cone shaped cutters mounted on roller bearings. This type of bit is the most common used in rotary drilling *(see drilling – rotary).* A multitude of tooth and cutter designs are available to suit different types of formations encountered. The choice of the correct bit to run in any situation is a vital decision as a soft formation bit will not attack a hard formation successfully any more than a hard formation bit will attack a soft formation. An error in the choice of bit type can be a very expensive mistake as a round trip *(see round trip)* to change a bit may take 20 hours or more and the cost, in the case of an offshore rig, will be in the region of £15 per minute.

connection: joining two lengths of pipe.

core: a solid bar of the formation being drilled which is recovered by using a 'core barrel'.

core barrel: *(see barrel – core).*

core bit: an annular type bit, which screws to the lower end of a core barrel and cuts a cylindrical bar of the formation which is retained in the core barrel by the 'core catcher' *(see core catcher).*

core catcher: a spring ring or fitting with spring loaded fingers which is located at the lower end of a core barrel and provides the means of retaining a core in the barrel.

coring (side wall): cores recovered from the wall of a hole already drilled and recovered by using an hydraulic or gun type device which forces small core retainers into the wall and thus obtains samples at that point for examination when the tool is pulled out of the hole.

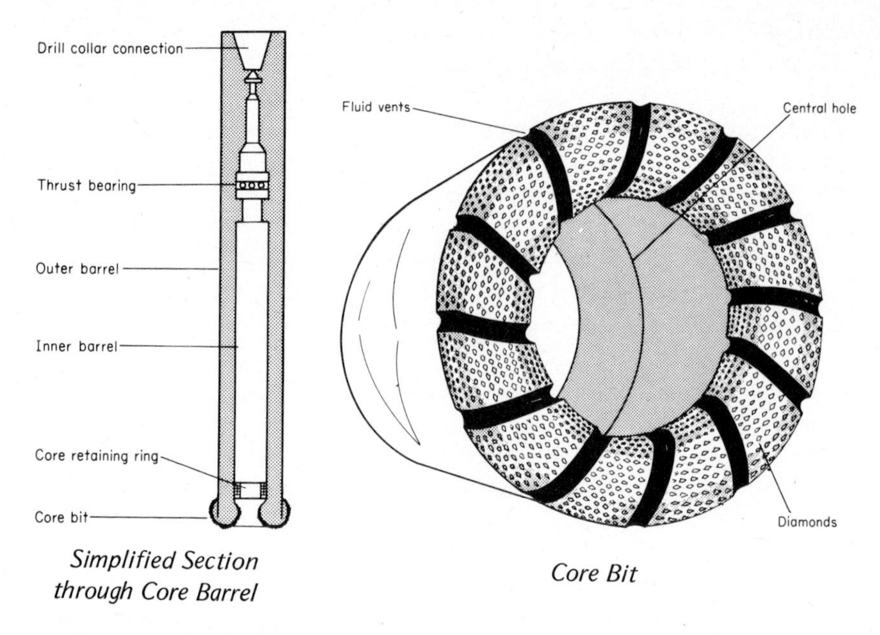

Simplified Section through Core Barrel

Core Bit

crack a valve: to open a valve slightly to allow a very small flow of fluid.

crown block: the sheave assembly at the top of a derrick or mast which accommodates the hoisting line from the drawworks *(see drawworks)* drum to the travelling block *(see travelling block).*

crude oil: oil from the formation in its crude form. The properties of crude oil from different fields vary tremendously. Specific gravity values vary, some oils are very heavy and have an asphaltic base whilst some have a wax content, others are light type oils or lubricating types of oil. The types vary from field to field according to the geological conditions which existed when it was laid down many millions of years ago *(see geology).*

cut oil: oil that contains water, also known as 'wet oil'.

cuttings: chippings from the formation being attacked by the bit and which return to surface in the drilling mud and are separated out by the vibrating screen *(see vibrating screen).* By examining the cuttings the geologist obtains information on the formations being penetrated and prepares his log *(see log – well).* The information so recorded from a series of wells in the same area provides valuable information on the down hole structures and influences the selection of future drilling sites or the decision to abandon operations in a particular area.

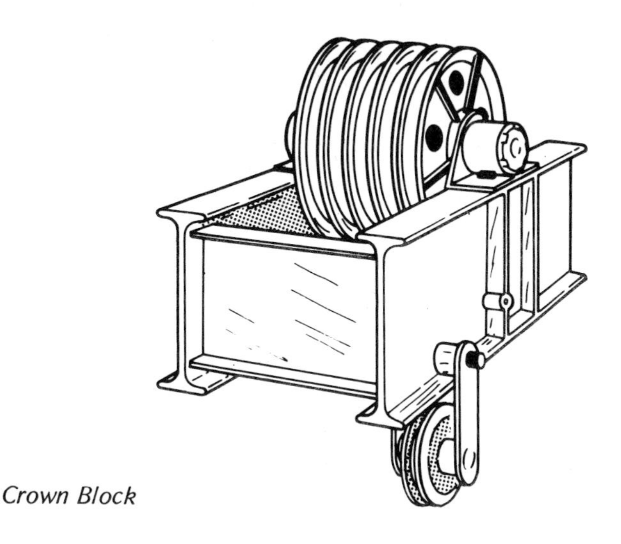

Crown Block

D

day tour: the shift period worked by a drilling crew from 0800 hours to 1600 hours.

deadline: the block line which is anchored to the 'dead line anchor' *(see anchor – dead line)* and extends from the floor to the 'dead sheave' of the crown block *(see crown block).*

deadline anchor: *(see anchor – deadline).*

decks: working platforms supported by the jacket *(see jacket)* of an offshore structure.

deck modules: steel box type containers in which production equipment, power units, meters and all types of equipment and machinery is installed for assembly on the deck areas of a production platform at sea. These units may weigh up to 1 700 tons (1 727 tonnes) and present major problems in transporting and loading on to the platform structures.

deflecting tool: wedge or other tool used for deflecting a hole from vertical.

deflection: deviation of a hole from true vertical.

degassing: removing gas from the formation oil by use of separators or similar plant at surface. The gas/oil ratio *(see gas/oil ratio)* varies tremendously from field to field with some samples containing little or no gas and some containing large quantities of gas.

Where there is a high gas/oil ratio in remote areas a problem exists as to disposal of the gas. The wasteful process of 'flaring' the gas is not normally allowed but where volumes are large the gas may be re-injected into the reservoir using wells specially drilled for this purpose.

derrick: the tower type structure erected over the spot which has been decided by the geologist is the site for a well to be drilled. The derrick is capable of handling all down hole operations and its capacity is dictated by the programmed depth of the well to be drilled and the drill pipe and casing loads to be handled. There are two main types of structure, one being the 'standard derrick' which is a tower type structure, or the more common 'drilling mast' which can be raised or lowered hydraulically and is relatively easy to transport from site to site. Either or each may be tested to withstand a pull of 1 500 000 lbs and is normally 136 ft (41 m) high in order to facilitate handling and stacking 90 ft (27 m) stands of drill pipe and casing.

derrick barges: large barges having a crane used for lifting heavy equipment and modules *(see modules)* on to an offshore platform *(see production platform).*

desander: a series of centrifuges used to remove sand particles from the circulating fluid on its return from the well and after it has passed over the vibrating screen *(see vibrating screen)* which removes the larger cuttings *(see cuttings).*

desilter: a series of centrifuges used to remove very fine particles from the circulating fluid on its return from the well and after it has passed over the vibrating screen *(see vibrating screen)* and through the desander *(see desander).*

die collar: a fishing tool *(see fishing tools)* used to recover drill pipe lost in a hole. The die collar is capable of cutting threads and is screwed on to the top of a 'fish' *(see fish)* usually by a string of drill pipe which is threaded with left hand threads.

differential pressure (bottom hole): the difference between the pressure existing at the bottom of the hole due to the fluid column in the hole and the flowing bottom hole pressure of the formation fluids.

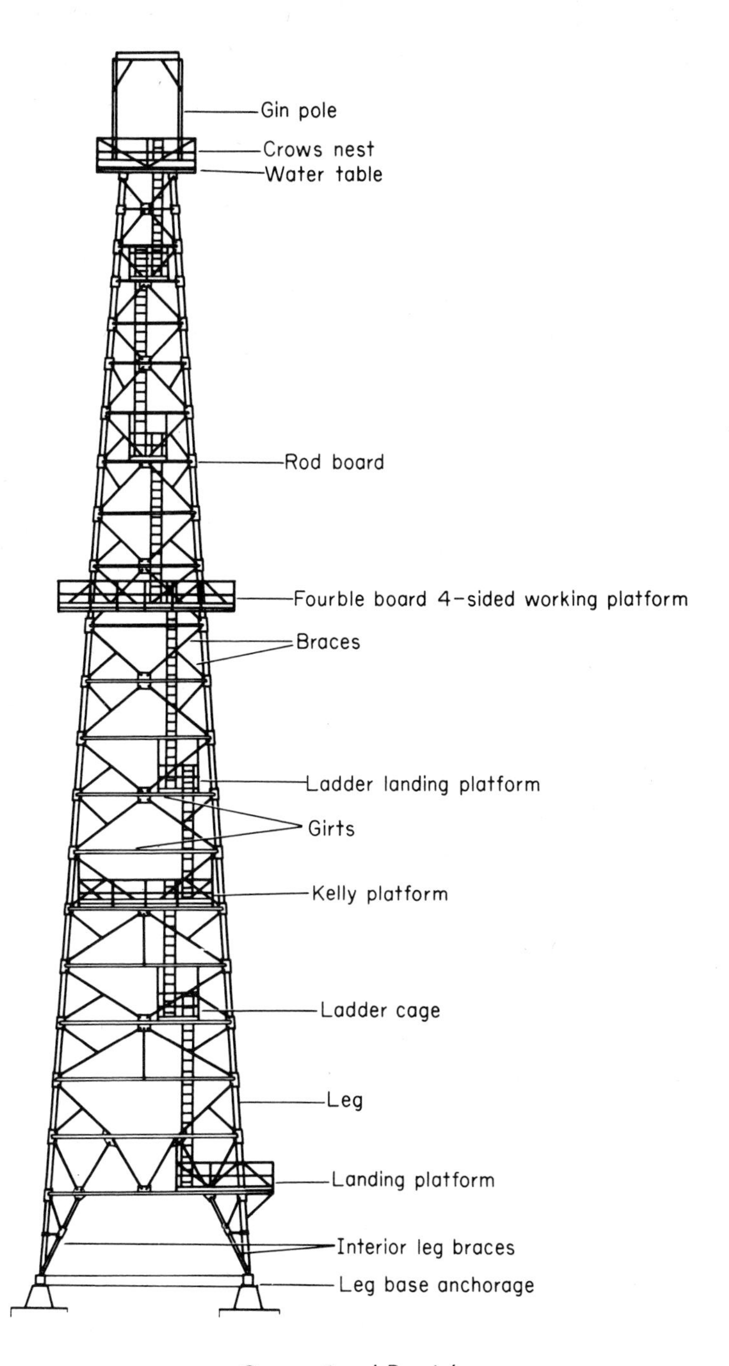

Conventional Derrick

dip meter: device for measuring the 'angle of dip' of a formation. From a geological point of view it is often important to know the inclination of a strata drilled through with relation to the horizontal. This information will assist in the siting of future wells. A 'dip meter' is run on an electric cable in a similar manner to other electric logs *(see log – electric)* and the information is recorded on a type of film in the recording unit at the surface.

directional drilling: controlled changing of direction of a hole from the true vertical. There may be many reasons for the need to drill a hole directionally:

(a) It may be impossible to site the rig *(see rig – drilling)* immediately over the target formation and the hole is then diverted to terminate at the desired point.

(b) It may be necessary to by-pass a 'fish' *(see fish)* and drill directionally in order to continue the vertical hole alongside the original hole.

(c) A 'wild well' *(see wild well)*, possibly on fire, may have to be killed by drilling into the bore at some depth and this can be done by erecting a rig, maybe a quarter of a mile distant, and directionally drilling into the original hole for the purpose of cementing up the 'wild well'.

(d) It is common practice to drill as many as 27 production wells directionally from a fixed platform at sea in order to produce a field. Such wells may deviate from 45° to 60° and their target points will include a circle of as much as two miles in diameter.

Though a tedious operation, wells can be drilled very accurately at controlled angles over long distances by means of using wedge-type 'whipstocks' *(see whipstock)* and down hole survey instruments, which record the angle and direction of deviation throughout the operation, making use of an inclinometer and camera.

dissolved gas: natural gas *(see gas – natural)* which is in solution with crude oil in a reservoir.

dog house: drillers' office on the derrick floor, which houses the 'lazy bench' *(see lazy bench)* and 'knowledge box' *(see knowledge box)*.

dog leg: a bend in the hole.

dome: a geological structure resembling a dome which if it has a cap rock *(see cap rock)* may contain oil or gas.

dope: lubricant of consistency as medium thick grease used on drill pipe and casing threads when making up a string of pipes.

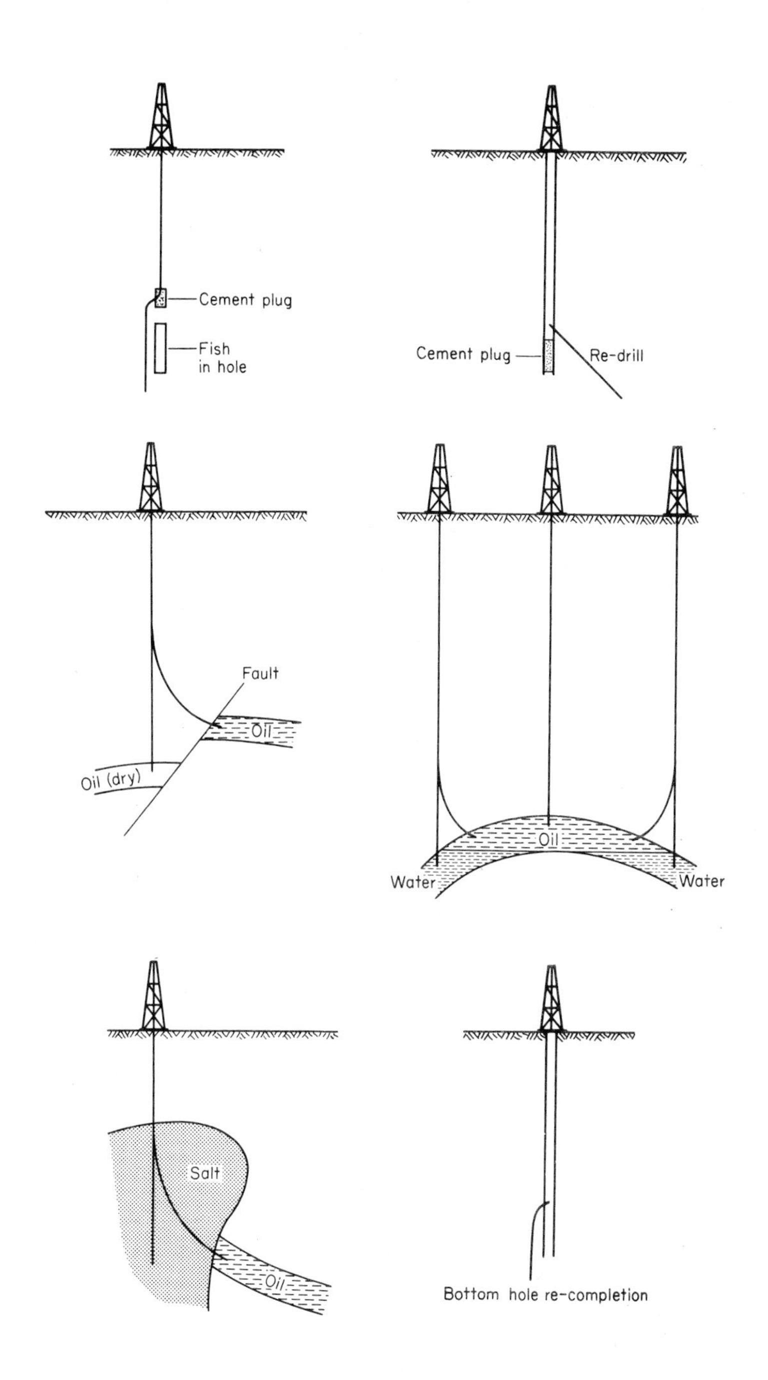

Directional Drilling

double: two joints of drill pipe screwed together forming a stand *(see stand)* approximately 40 ft to 60 ft (12.2 m to 18.3 m) in length.

double board: a platform erected in the derrick to allow the derrick man to handle 'doubles' *(see double).*

double box: a connector screwed double female with fast tapered threads to accommodate two drill pipe pins *(see pin).*

double pin: a connector screwed with fast tapered double male threads to accommodate two drill pipe boxes *(see tool joint).*

doughnut: a ring of wedges fitted with slips *(see slips)* that supports a string of drill pipe, casing, or tubing.

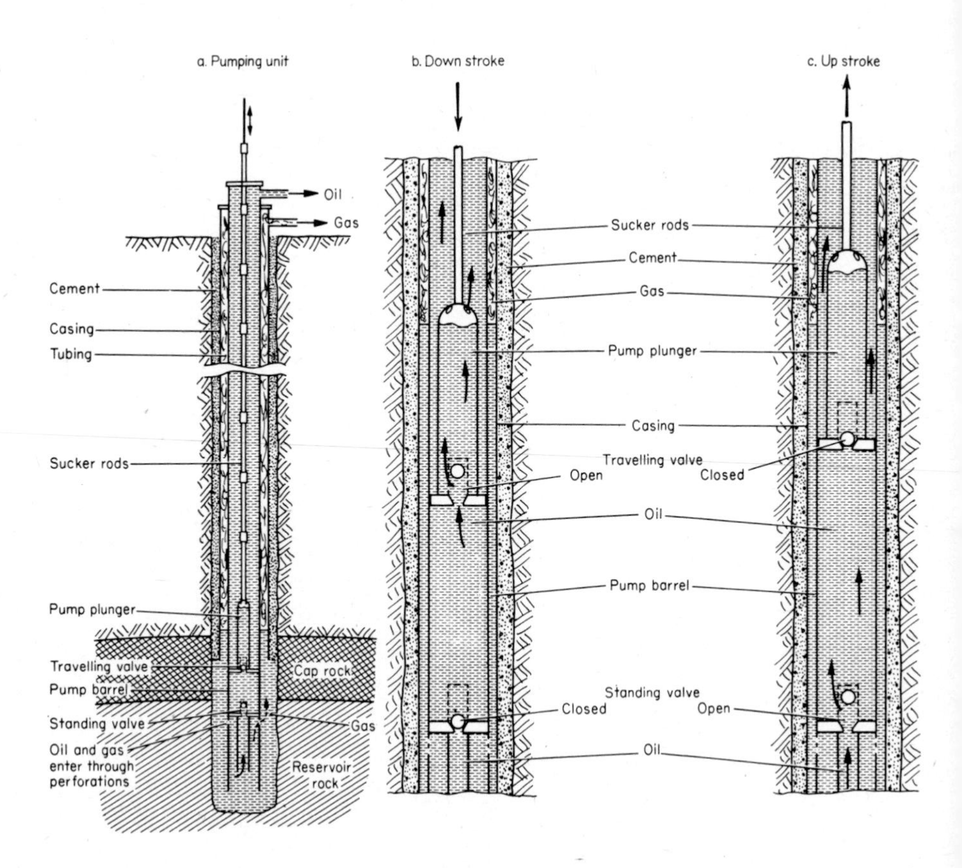

Down hole Pump

down hole pump: a pump which is installed in a non-flowing well below the oil level and is operated by a pumping jack *(see pumping jack)* or a down hole electric motor for the purpose of producing the well.

down hole turbine: *(see turbo drill).*

drag bit: a blade type bit.

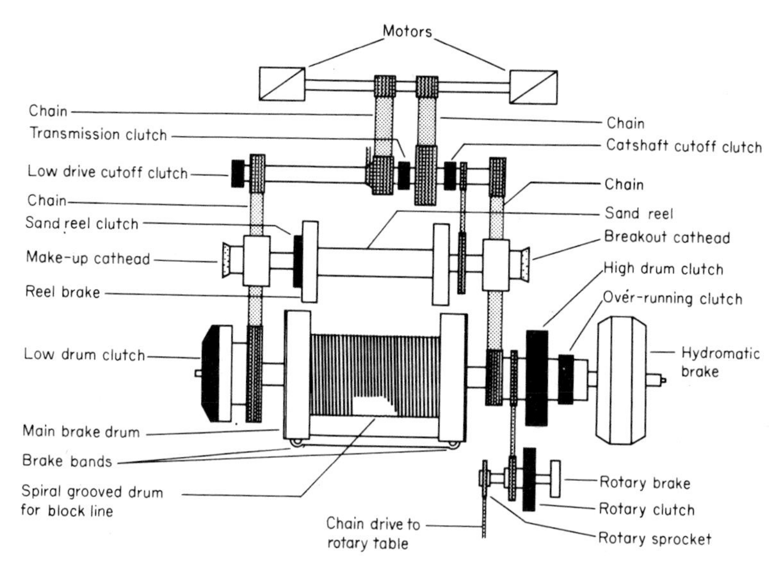

Drawworks Layout

drawworks: the main winch mounted on the derrick floor which handles the entire power for raising and lowering the drill pipe and casing loads and supplies power and controls for rotating the drill pipe via the rotary table *(see rotary table).* To handle a deep well operation the drawworks will require an input horse power of some 3 000 H.P.

drill collar: heavy sections of pipe about 30 ft (9.1 m) long which are run in the drill string immediately above the rock bit to provide weight on the bit and also to hold the drilling string above in tension to assist in drilling a vertical hole. A drill collar string also provides a fly wheel effect to absorb shock loads due to the drilling operation and relieves the more fragile drill pipe of torsional loads. Common drill collar sizes are 10 inch OD, 7 inch OD, 6 inch OD, 4⅝ inch OD (25 cm, 18 cm, 15 cm, 12 cm).

The diameter of the drill collars used is, of course, dictated by the size of hole being drilled as the collars must be a little smaller than the bit diameter and should permit the running of any 'overshot' *(see overshot)* if the collars should become stuck in the hole.

drill pipe: high grade pipe which screwed together makes up the drill string used in drilling a well. Singles of drill pipe are approximately 30 ft (9.1 m) in length and are fitted at the lower end with a pin connection (tool joint) and at the upper end with a box connection (tool joint).

Some of the most common pipes in use are: 5 inch OD, 4½ inch OD, 3½ inch OD, 2⅞ inch OD, 2⅛ inch OD (13 cm, 11.4 cm, 8.9 cm, 7.3 cm, 5.4 cm). The most common of these is the 4½ inch OD (11.4 cm) string.

drill ship: floating unit fitted with a drilling rig *(see rig – drilling)* which can be moved from location to location as a ship and can be anchored at the drilling site or held on station by computer control.

The drill ship is a useful tool where water depths are too great for a fixed platform to be erected either on account of economic or technical reasons. Drill ships can handle operations in great depths of water – 2 000 ft (610 m) or more – but they present problems on maintaining station in storm conditions and on account of the fact that they rise and fall with the wave motion. Consequently when using a 'drill ship' provision has to be made to set the well head and blow out preventer stack *(see blow out preventer stack)* on the sea floor and to provide a compensating riser pipe from the well head to under the derrick floor which is capable by a telescopic and flexible type joint to accommodate the motion of the ship.

In order to maintain weight on the rock bit *(see rock bit)* when drilling, it is also necessary to include expansion joints in the drill string *(see drill string)* to prevent the rise and fall of the ship from lifting the drill collar string *(see drill collar)* and thus remove the load from the bit. Additional problems are also introduced in maintaining the blow out preventer stack at depth and the need to employ expensive divers to assist in the related underwater operations at great depths. Power for the rig machinery and pumps on a drill ship is generated by the main propulsion engines and the drawworks and pumps are usually powered by electric motors *(see also floating rig)*.

In the North Sea where water depths do not normally exceed 600 ft (183 m) it is preferable to make use of semi-submersible units *(see semi-submersible rig)* which are more stable in rough conditions than a 'drill ship'. The drill ship, however, has the advantage of being able to travel from location to location under its own power much more readily than a 'semi-submersible' and is, therefore, more adaptable to long moves.

drill stem test: a method of testing the potential production from a reservoir formation by running a packer *(see packer)* on a drill string to accommodate the hydrostatic head of the mud column and allow the formation fluid to flow into the drill pipe. The test tools allow means of opening and closing the device from surface and incorporate a pressure recorder which gives permanent information on initial formation pressure, flow records, and build up pressure values when the tool is finally closed. The information provided by the flow results and recorder chart will allow the petroleum engineer to assess the potential production rate of a reservoir, and results from a series of wells will provide a reliable estimate of the capacity of a 'field'.

drill string: the column of drill pipe made up of singles *(see single)* and drill collars *(see drill collar)* which extends from surface to the bottom of the hole and provides the means of rotating the rock bit and circulating the mud *(see mud).*

drilling: the operation of boring a hole in the earth's crust for the production of hydrocarbons, steam, or water. A hole may also be drilled to obtain geological information by examination of the formation cuttings returned to the surface in the drilling mud or from rock cores obtained by using special coring equipment *(see barrel – core).*

drilling (cable tool): the method of drilling a hole using a chisel-type bit suspended on a wire line and reciprocated to pulverise the rock formation for removal by a bailer *(see bailer)* or sand pump *(see sand pump).*

NOTE. This method was used extensively until the general introduction of 'rotary drilling' *(see drilling – rotary)* around the early 1930s. Cable tool drilling is, to this day, extensively used in the drilling of water wells. The method is very slow in comparison with rotary drilling and a serious disadvantage is that there is no fluid column in the hole to control any pressure shows which may be encountered. This factor accounted for the 'gushers' which were common in the early days of exploration.

drilling (diamond): diamond drilling rigs are commonly used for mineral investigation when cores of the formation are required for analysis. The rotary system *(see drilling – rotary)* is used but the bits are set with industrial diamonds and rotary speeds are much higher (up to 2 000 to 3 000 r.p.m.) than those used in oil well drilling practice using conventional rock bits *(see rock bit)* where rotary speeds above 250 r.p.m. are unusual.

Diamond bits are frequently used for directional drilling of oil wells and in these cases are often driven by a 'down hole turbine' *(see turbo drill)* powered by the circulating fluid serving the well *(see mud).*

drilling (percussion): one form of percussion drilling is the cable tool method *(see drilling – cable tool).*

However, percussion hammers are also used with the rotary system in situations where compressed air circulation is possible when formation fluid is not

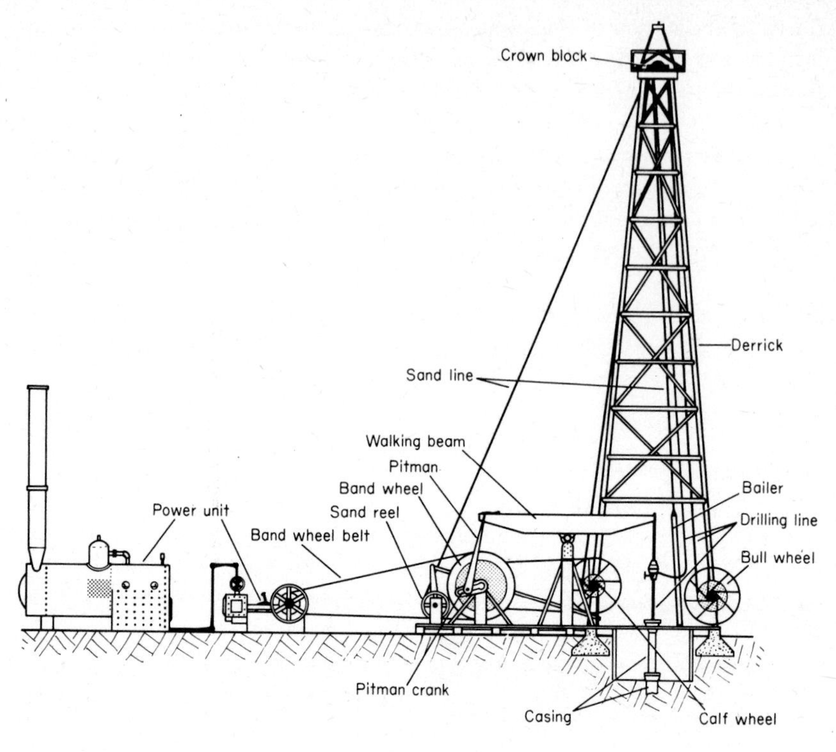

Drilling (Cable tool)

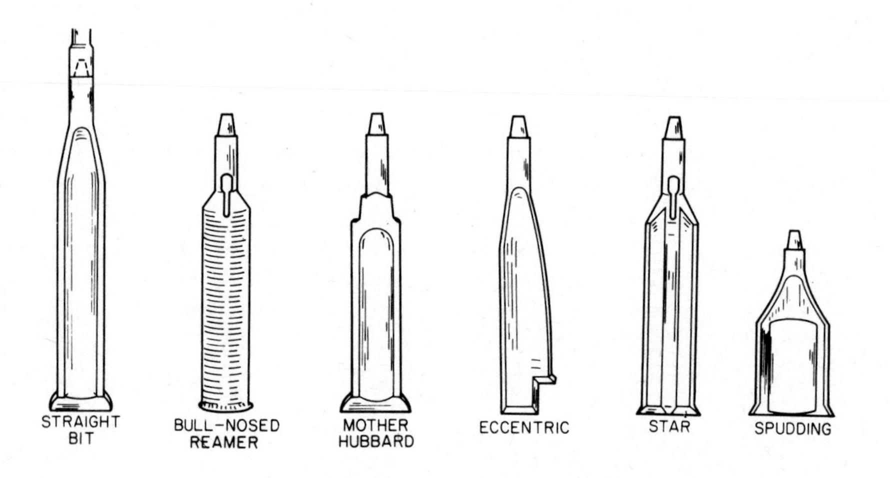

Cable tool bits used in cable tool drilling

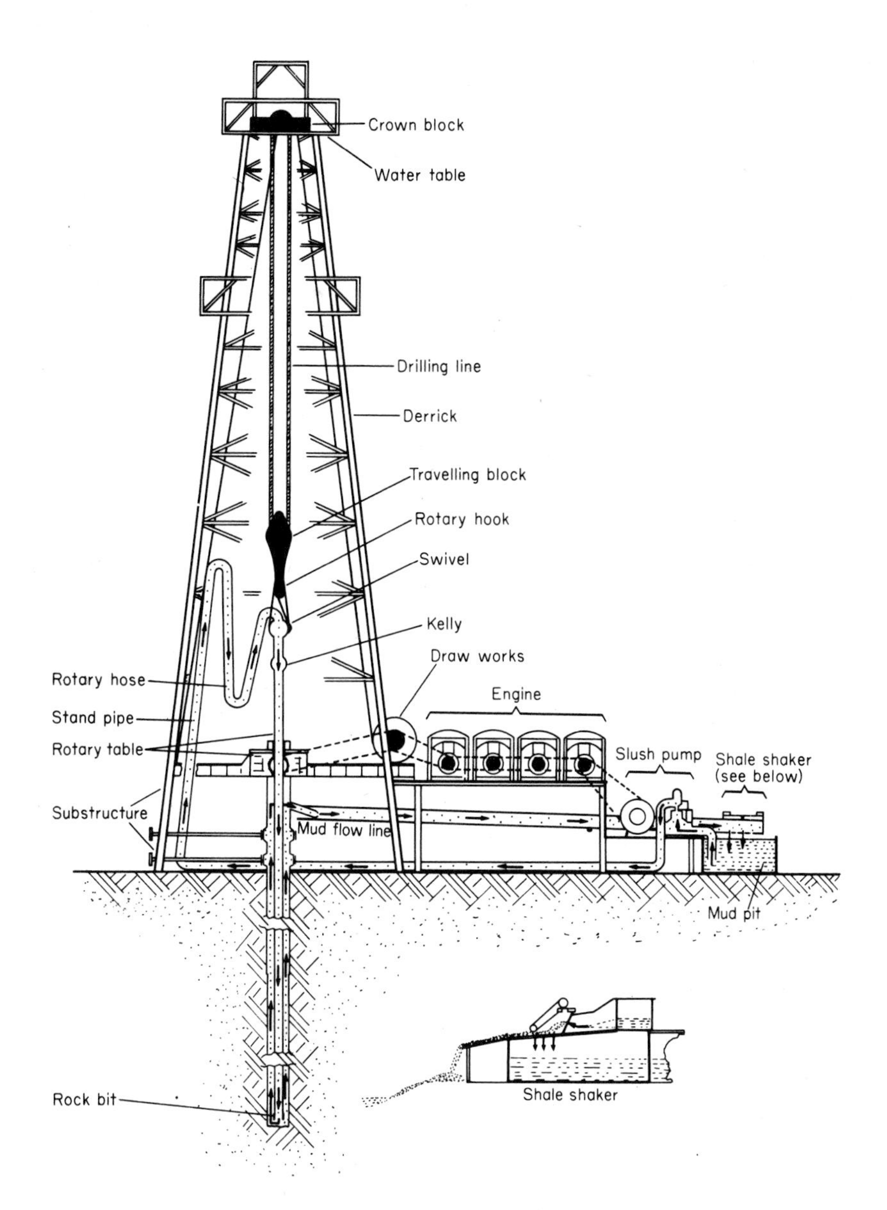

Rotary Drilling

present to flow into the hole. Penetration rates and bit life are much improved by the use of air hammers and this method is much more commonly used in mineral investigation projects than when drilling for oil or gas. Air circulation is not possible when fluid is flowing into the hole from a formation in any quantity as the fluid column generated causes a back pressure which prevents the air circulation and the operation of the hammer. Attempts are being made to introduce a down hole hammer which can be operated by the fluid circulating in a well instead of by the use of compressed air.

drilling (rotary): all modern oil well drilling uses the rotary system whereby a rock bit *(see rock bit)* is rotated by a string of drill pipe extending from the surface to the bottom of the hole. Cuttings from the hole are removed to surface by circulating 'mud' *(see mud)*, water or air through the drill pipe, discharging it through nozzles in the bit and returning to surface in the annular space between the drill pipe and the well wall.

drilling crew: from the 'spudding in' of a well *(see spud in)* to its completion the rig *(see rig – drilling)* operates on a 24 hour a day basis. Three shifts or 'tours' are run, usually of an eight hour period each except on some offshore operations when two twelve hour 'tours' are run. Relief crews take over to provide rest periods.

In charge of the drilling operation is the **Drilling Superintendent.** Second-in-command is the **Tool Pusher** who is responsible to the Superintendent for the day-to-day operations and for ensuring that all necessary equipment, tools and materials and services are available as required.

The Driller is responsible for his crew and the running of the rig during his eight or twelve hour 'tour' and he is responsible in line to the Tool Pusher. A **Trainee Driller** may be included in the crew.

The Derrick Man is the next most important member of the crew and his duties are mainly concerned with the handling and racking of drill pipe stands *(see stand)* as they are pulled or run during a round trip *(see round trip).* His place of work is at the fourble board *(see board – fourble)* some 90 ft (27.4 m) above the derrick floor.

Rough Necks work on the derrick floor and handle the placing of the 'slips' *(see slips)* in the rotary table *(see rotary table)* to support the drill string on a round trip and to handle the make-up and break-out tongs and the spinning chain *(see spinning chain)* which are used to make or break joints when running or pulling a drill string.

Roustabouts are members of the crew who handle the loading and unloading of equipment and assist in general operations around the rig site.

drilling in: the term used when a well is drilled into the producing formation.

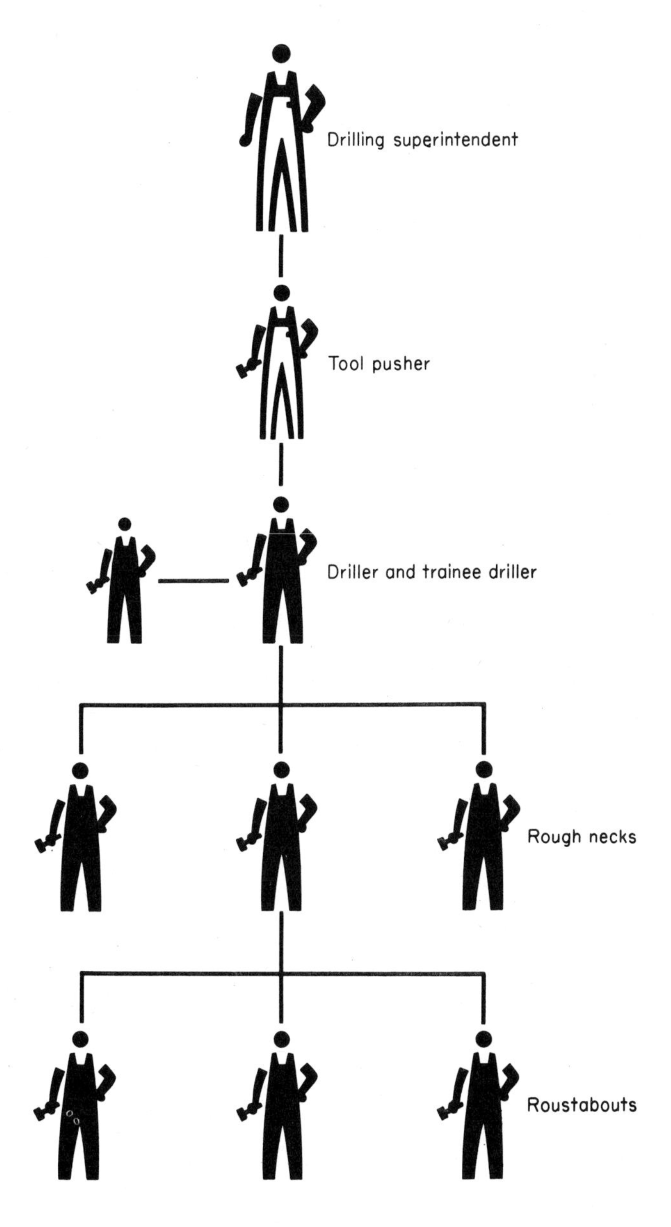

Drilling Crew

drilling log: *(see log – drilling).*

drilling mud: *(see mud).*

drilling out: the term used for drilling residue cement at the bottom of a casing string *(see casing string)* after a cement job *(see cement job).*

drilling spool: a spacer fitted into the wellhead and usually 'double flanged'. A spool may serve to act as an adaptor for connecting two items with unlike flanges in a case such as when a blow out preventer *(see blow out preventer stack)* having a given size flange must be matched to a casing string which is fitted with a different size flange.

drive shoe: *(see shoe – drive).*

dry hole: non-productive hole also known as a 'duster'.

dual completion: a well which will produce from two separate reservoir formations at different levels and which is provided with two strings of production tubing and a packer assembly to isolate each producing formation from the other. Some wells have as many as five separate production tubings set in this manner.

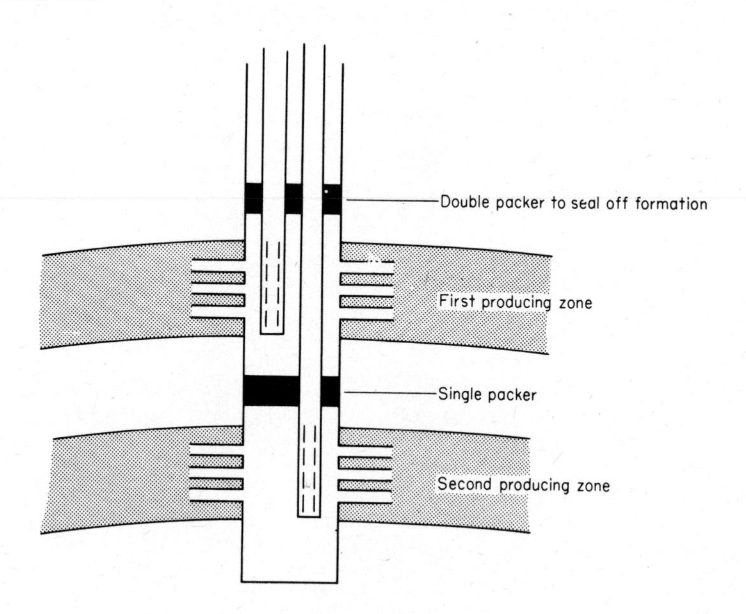

Dual Completion

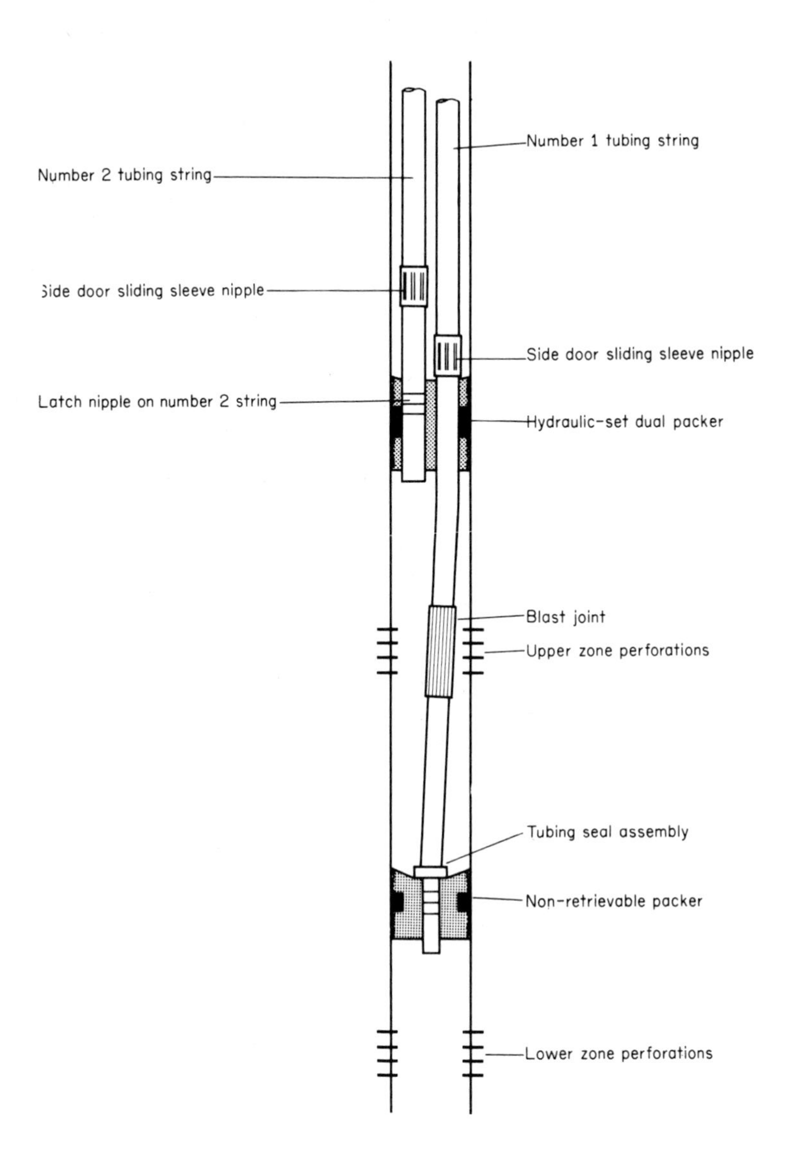

Dual Completion

Duplex pump: a two cylinder double acting pump with interchangeable liners and pistons used for handling the circulating mud *(see mud)* or fluid when drilling a well. These pumps can accommodate varying pressure and volume situations as required by well conditions which change as the hole progresses from surface to its final depth.

The pumps installed at the start of a well are matched to the volume and pressure requirements for the particular well casing programme *(see casing string)* as designed.

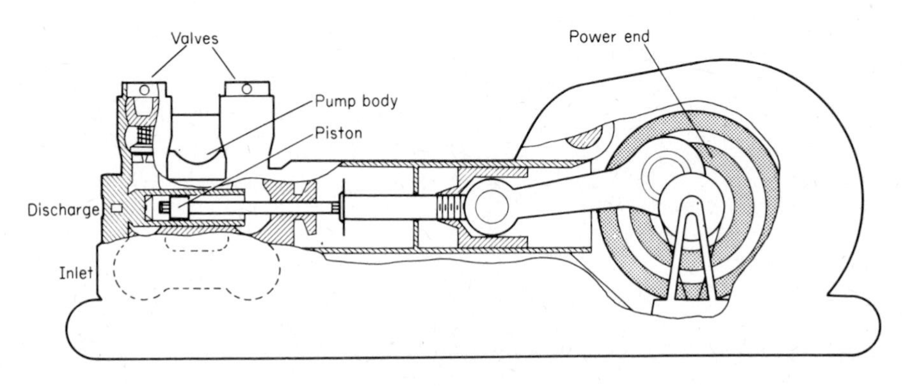

Duplex Pump

A typical Duplex pump powered by a 1 600 BHP engine would have the following volume and pressure capacities:

Liner size (inch)	7¼	7	6¾	6½	6¼	6
Maximum discharge pressure (lbs/in^2)	3515	3810	4135	4505	4940	5440
Volume GPM discharge	665	615	565	515	470	430

Liners and pistons are changed during the life of the drilling operation to accommodate the changing volume pressure condition requirements.

duster: a well which has proved to be non-productive or a 'dry hole'.

Dutchman: the part of a stud or screw which remains in place after 'twisting off'. Also the portion of a tool joint pin *(see pin)* which is left in a box *(see tool joint)* after 'twisting off'.

edge water: water in a formation on the outer rim of an oil reservoir which in certain circumstances provides 'water drive' *(see water drive)* to produce the field. This situation is common in limestone reservoirs such as are found in some of the big Middle East oil fields.

electric logging: the method of surveying a bore hole by means of running special tools on an electric cable which records on tape the identity of formation properties in a well and can provide details of rock types, fluid content, porosity *(see porosity)*, permeability *(see permeability)*, dip *(see dip meter)* and many other details of down hole conditions.

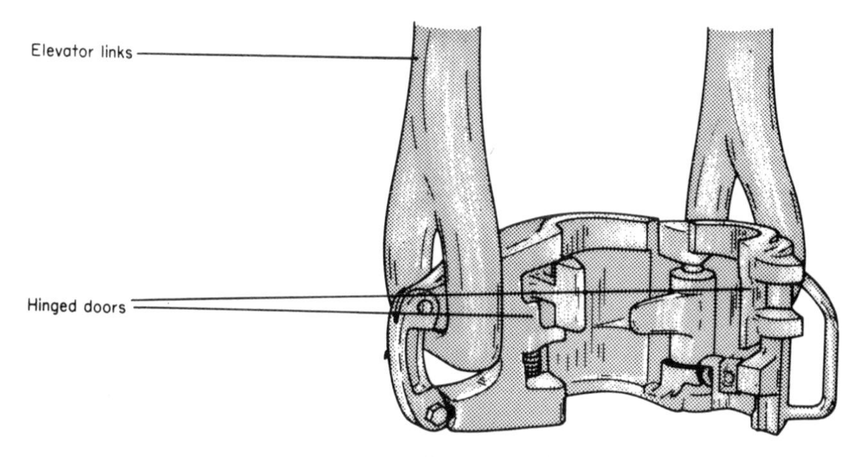

Elevators (Collar type)

elevator: a lifting device with hinged doors and a fast releasing latch which hangs on long links below the travelling block and hook and which, in a closed position, fits snugly around the drill pipe or casing to handle lifting and lowering of a string of drill pipe or casing.

elevator links: long steel links which connect the elevators *(see elevator)* to the main hook.

exploitation well: a well drilled to explore the potential of a field *(see field)*. A number of such wells are drilled after the original discovery well in order to establish the extent of a field and to provide information from which estimates of the reservoir capacity and its production potential can be arrived at.

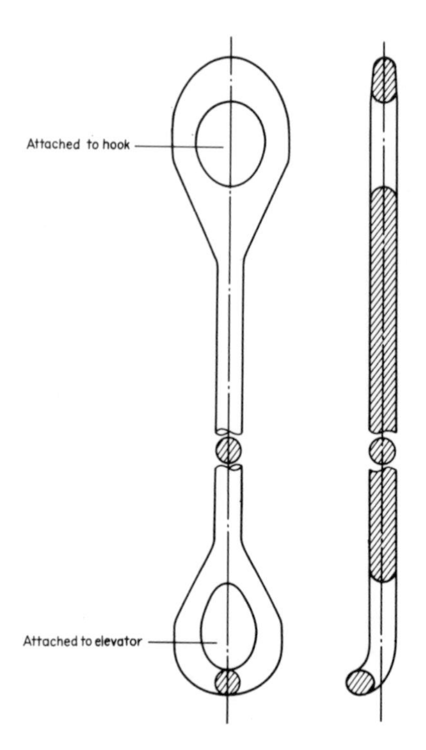

Elevator Link

F

fast line: the line which spools on to the main drum of the drawworks *(see drawworks)* and which derives its name from the fact that it runs faster than the lines spooled in the travelling blocks *(see travelling block).*

fault: a geological term for a break in the subsurface formation.

feed off: running the drill string continuously whilst drilling without using the drawworks *(see drawworks)* brake. Down hole conditions only allow this to be done when soft formations are encountered and steady penetration rates of the formation being drilled allow a regular 'feed off' rate.

field: an area consisting of a single reservoir or multiple reservoirs all grouped on, or related to, the same individual geological structural features and/or strata graphic conditions. Usually a 'field' is only declared when oil has been discovered in sufficient quantities to justify a profitable production programme.

filling the hole: pumping mud into a well to replace the volume of drill pipe being pulled out when round tripping *(see round trip)* or pumping mud to keep the hole full when a thief formation *(see lose returns)* is taking fluid.

It is at all times necessary to keep the hole full of fluid in order to ensure that fluid or gas under pressure in a formation cannot flow into the well in an uncontrolled manner and thus present a hazard.

filter cake: material remaining on a filter paper when the mud properties are tested with a 'filter press' under 100 lbs/in^2 (7 kg/cm^2) air pressure.

finger: a pipe or rack fixed in the derrick *(see derrick)* at some 90 ft (27 m) above the floor for the purpose of racking stands *(see stand)* of drill pipe when round tripping *(see round trip).*

fish: any tool or part of a drilling string *(see drill string)* or similar item which is lost in a hole and involves a 'fishing job' before normal operations can be recommenced.

fishing hook: a fishing tool *(see fishing tools)* used for pulling the top of a 'fish' *(see fish)* into the centre of the hole to allow engagement of a recovery tool.

fishing jar: a fishing tool *(see fishing tools)* used to exert an upward blow to a stuck pipe.

fishing magnet: a permanent type of magnet used to recover small steel items from a hole such as cutters or bearings from a collapsed bit.

fishing tap: a long tapered tap used to screw into the top of a 'fish' *(see fish)* and often run in conjunction with a set of fishing jars *(see fishing jar).*

fishing tools: numerous devices for recovering a fish *(see fish).*

fishtail bit: blade type bit suitable for attacking soft formations.

float collar: a short length of casing screwed to the lower end of a casing string *(see casing string)* and fitted with a non-return valve which provides a means of floating a casing string into the hole and thus relieving the hoisting gear of excessive loads.

float shoe: casing shoe *(see casing shoe)* fitted with a non-return valve which operates in a similar manner to a 'float collar' *(see float collar).*

floating rig: a floating rig or a 'floater' is a ship type of vessel fitted with drilling equipment capable of drilling in deep water – 2 000 ft (610 m) or more – and usually having a drawworks *(see drawworks)* capacity to drill to 25 000 ft to 30 000 ft (7 620 m to 9 144 m). These vessels are usually self-propelled for moving from one site to another and are anchored on station when drilling or held on station by computer control operated thrusters.

flow bean: a choke or adjustable choke used to restrict the flow of fluid or gas from a production well.

flow line: return pipe from well head through which mud or fluid passes to the mud screens *(see vibrating screens)*. Or a pipe through which a well produces oil or gas.

flow string: the final tubing or casing set in a production well.

flow tank: a storage tank into which a producing well discharges.

flow test: controlled production from a well to establish reservoir conditions.

flowing well: a well which produces oil or gas without any means of artificial lift.

fluid level: the level of oil from the well head to which fluid rises in the bore in a well which is not a flowing well *(see flowing well)*.

flush joint casing: casing joints which have a regular outside diameter as opposed to collar casing where the connecting collars have a larger outside diameter than that of the casing joints.

formation: *(see geology)*.

formation pressure: the pressure at the bottom of a well when it is shut in at the well head.

formation tester: a packer assembly which is set in the open hole or in a casing above the producing formation whereby the hydrostatic head of the mud column is contained and this allows fluid from the formation to flow freely into the test string and to surface where it can be measured and analysed. The formation test tool usually includes a down hole recording device which supplies vital information on the pressure conditions existing in the reservoir rock and its flow potential.

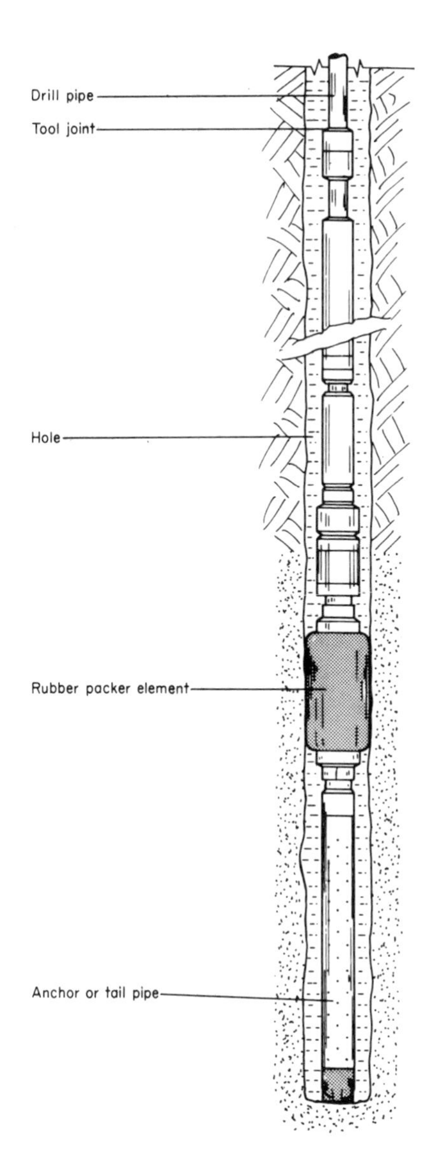

Formation Tester

fourble: four joints of pipe screwed together to form a 'stand' *(see stand).*

fourble board: *(see board – fourble).*

fracturing (formation): the method of breaking down a down hole formation by pumping very high pressures to increase production of a reservoir

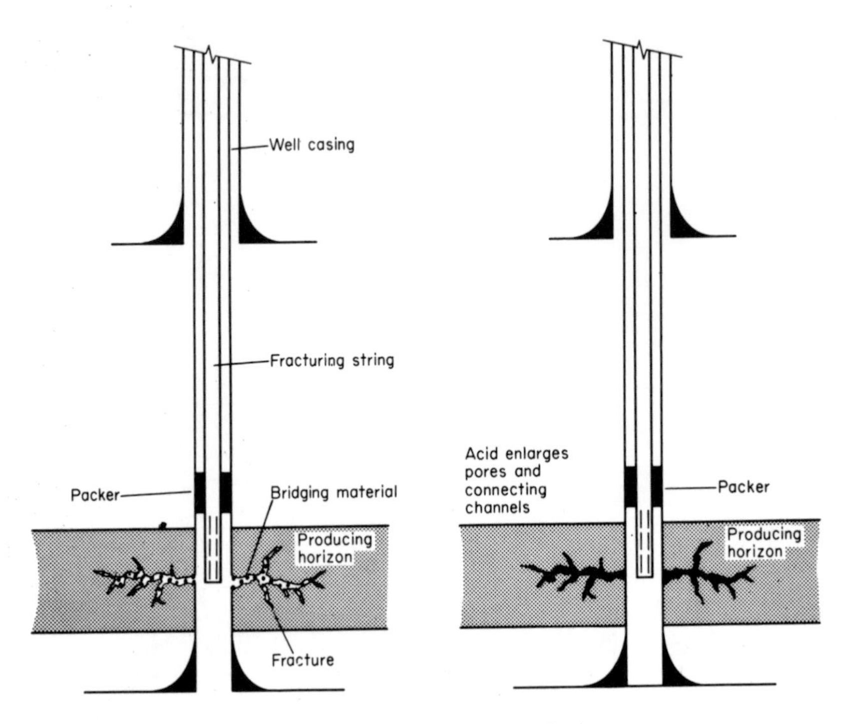

Fracturing (Formation)

formation. By this technique the formation is 'cracked' and the fissures are propped open by pumping walnut shells, glass balls or similar propping agents into the formation.

free point indicator: an instrument which is run into a stuck drill pipe string on an electric cable and which records the point at which the string is stuck.

friction socket: a 'fishing tool' *(see fishing tools)* used for recovering drill pipe which has been lost in the hole. The tool is a tube which can be forced over a 'fish' *(see fish)* and will thereby take a friction hold to recover pipe which is not firmly stuck in the well. Little used nowadays as it is a relic of the old 'cable tool days' *(see drilling – cable tool)* and much more sophisticated tools are now available.

frost up: icing up of equipment due to the expansion of gas when it passes from a situation of high pressure to a situation of lower pressure through a partially closed valve or a choke or some similar situation.

full hole: a bore hole drilled to full diameter.

full hole tool joint: screwed connector used for joining joints of drill pipe and having the same bore as the pipe itself.

G

gas (natural): unprocessed gas from a gas reservoir.

gas (stripped): processed natural gas with condensate removed. The condensate is removed by passing high pressure gas through a series of separators (cylindrical pressure vessels) and reducing the pressure in stages. The gas then cools due to expansion, and condensate is produced and can be drawn off fron the bottom of the separator.

gas (wet): natural gas containing condensate. The amount of condensate varies from well to well and may be 'very wet' or almost 'dry'. A wet gas gives rise to hopes that the gas is being recovered from the vicinity of an oil reservoir.

gas cut mud: mud in circulation *(see mud)* which contains entrained gas which has seeped into the column from a gas reservoir. When the mud column only contains a little gas it can be removed by gunning *(see gun – mud)* vigorously in a tank before being returned to the well. The situation can, however, be serious when the column is heavily gas cut as this condition reduces the hydrostatic pressure in the well and may allow the well to 'come in' *(see come in)*. In such a case it becomes necessary to shut down drilling operations to close the well head blow out preventers *(see blow out preventer stack)* and to replace the mud in the hole by pumping in a new volume of good mud fron storage.

gas drive: pressure exerted by formation or injected gas to cause or assist a production flow.

gas lift: gas flowed into a well through a pipe or tubing in order to lighten the fluid column and induce a production flow from the reservoir.

gas oil ratio: the number of cubic feet of gas produced from a barrel of oil at atmospheric pressure.

gas well: a well which produces natural gas.

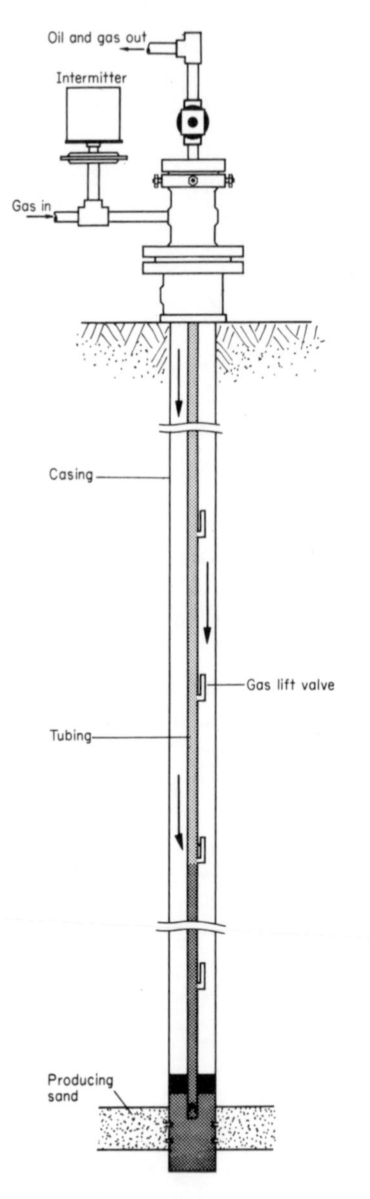

Gas Lift

gel: the ability of mud to hold solids in suspension when normal circulation by pumping is suspended. This property prevents cuttings in the mud column from settling to the bottom of the hole and sticking the drill string if and when normal circulation is stopped.

geolograph: an instrument which records the speed of penetration of the bit during drilling operations. The rate of penetration will depend on a number of factors and they may vary between a few inches an hour to a rate which is only limited by the ability of the driller to make a connection *(see making a connection)* i.e. add 30 ft (9.1 m) lengths of pipe to the drill string. Some factors affecting the penetration rate may be:

(a) type and hardness of the formation;
(b) type of rockbit in use;
(c) type of mud in circulation;
(d) weight applied to the bit by the drill collar string;
(e) speed of rotation of the bit to suit the formation;
(f) expertise of the driller;
(g) capacity of the slush pumps to handle the circulating fluid.

geology: oil and gas is discovered in formations of the earth which were laid down many many millions of years ago and, in some cases, as much as 350 million years. The oil is normally found in what are known as reservoir rocks which are not usually the formations in which it was initially formed, which are known as 'source rocks'. The oil, of course, is not in place as a kind of lake but is contained in porous rock such as limestone or sandstone something like water may be contained in a common sponge. It must be realised that conditions on earth at the time we are considering were entirely different from conditions at the present time. Rain fell in torrents for thousands of years at a time and the earth's surface was subjected to tremendous movement, volcanic eruptions, earthquakes and extremes in temperature. The lakes and seas teemed with all kinds of strange life and this was the age of the huge reptiles like the dinosaurs.

It is generally accepted that oil and gas were formed in sedimentary formations i.e. strata which is formed from material which was washed down from the mountains by the great rains and settled as deposits of silt and mud where the rivers ran into the lakes or seas. This process went on for millions of years and the deposits became hundreds or thousands of feet thick. Into these soft deposits fell dead bodies of the teeming life and 'plankton' which existed at that time. In later ages further materials such as sands, limestones and shales covered the original deposits again to depths of hundreds or thousands of feet.

Over the ages cooling of the earth's crust caused tremendous pressures to be exerted on the compacted mass of deposits and their contents and the formations were folded into mountains and valleys and formed 'anticlines' and 'synclines' or huge ridge type structures.

The tremendous heat and pressures caused the fatty materials of the dead organisms to generate petroleum in countless millions of tiny globules. The oil tended to move upwards as it was displaced by water and almost certainly

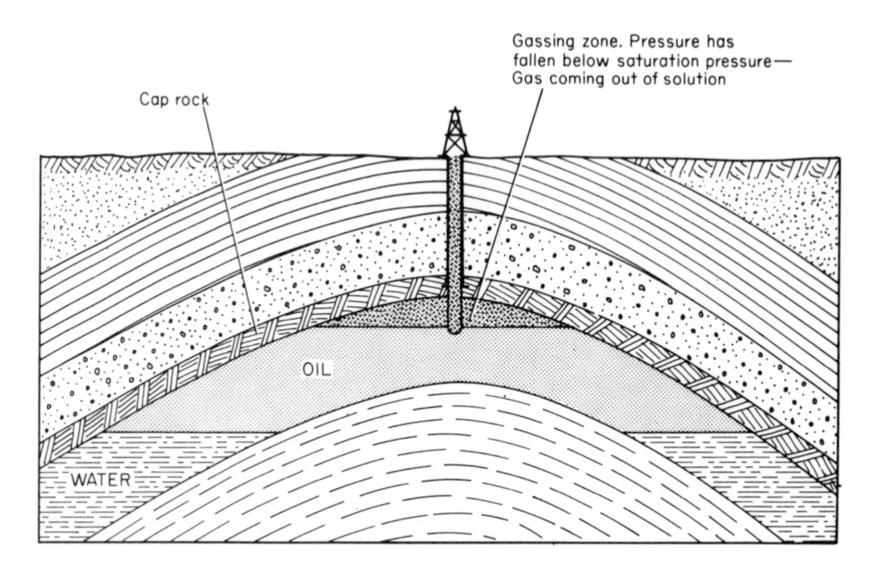

Geology: Sedimentary formations

most of the oil reached the surface of the earth and was lost for ever to the atmosphere. However, some of the oil became entrapped in dome like structures caused by the movement of the earth's mantle in cases when an impervious layer of rock, known as cap rock *(see cap rock)* overlay these 'domes'. In their search for oil deposits the geologists and the geophysicists look for these dome like structures of the rock formation and an age period which could possibly contain oil. There is no means at present of knowing whether such a formation will contain oil until a very expensive well has been drilled.

There is a very wide range of types of crude oil from different fields and this is understandable when one takes into consideration the changing conditions of climate and life which existed and gave rise to its formation over the many millions of years during which it was produced.

It must be realised, of course, that the above is only a very sketchy description of the theory behind the formation of oil, but it will assist in clarifying some aspects for those who have no previous knowledge of the subject.

girth or girt: one of the horizontal braces between the main legs of a derrick.

gone to water: a well producing edge water *(see edge water)* as production from the field affects the gas/oil water levels. The well concerned must then be closed down and production taken from other wells closer to the centre of the field. Eventually, of course, all wells in a field will 'go to water' in a field which depends upon 'water drive' *(see water drive)* such as is common in limestone formations.

goose neck: connecting bend between the 'rotary swivel' *(see swivel)* and the 'kelly hose' *(see hose – rotary).*

gravel packed completion: in cases where a reservoir formation is made up of loose fine grained sand, which tends to fall into the hole, it is common practice to set a slotted liner fitted with a wire screen and to pack the annular space between the well wall and the liner with gravel, which acts as a filter to prevent the sand from entering the well with the produced fluid. The gravel pack is placed in position by pumping it in a fluid, such as mud, down the annular space between the well wall and the final casing string whereby the gravel remains outside the wire screen and the mud flows through the screen and returns to surface.

graveyard tour: the drilling shift which normally commences at midnight and terminates at 0800 hours.

gravity (specific): the ratio of the weight of a volume of substance being considered to a similar volume of water at NTP. In the case of a gaseous material the standard against which the gas is measured is 'air'. The formula is:

$$\text{Degrees A.P.I. gravity} = \frac{141.5}{\text{specific gravity}} - 131.5$$

gravity A.P.I.: throughout the U.S.A. and many other parts of the world the A.P.I. gravity scale is used. The relationship between specific gravity and the A.P.I. scale is:

$$\text{Specific gravity} = \frac{141.5}{131.5} + {}^{\circ}\text{A.P.I.}$$

Thus it will be seen that as the specific gravity increases the A.P.I. value decreases, e.g.:

Heavy lubricating oil	s.g. .966	would be A.P.I. 15
Petroleum spirit	s.g. .8017	would be A.P.I. 45
Aviation spirit	s.g. .6690	would be A.P.I. 80

grief stem: the kelly *(see kelly).*

guide shoe: short section of thick wall pipe screwed to the bottom end of a casing string *(see casing string)* to protect the casing when running into the hole.

gun (mud): a stand pipe fitted with a swivelling jet pipe which is used for agitating fluid in a tank for the purpose of thoroughly mixing the ingredients of a drilling mud.

gun perforating: device for perforating casing at a point opposite a reservoir formation to permit the production of fluid from the formation into the flow string. The gun which fires bullets sideways is run into the casing on an electric cable and when in the desired position is fired from the control point at surface.

gusher: a flowing well, possibly not under control.

H

head board: *(see board – head).*

headache post: a post on the floor of a 'cable tool' rig *(see drilling – cable tool)* which prevents the 'walking beam' *(see walking beam)* from falling when the 'pitman' *(see pitman)* is disconnected from the driving crank.

heaving shale: shale formation which tends to crumble, swell or fall into the hole in a persistent manner.

high line: combination of a rope and wire line which is used for pulling pipe and equipment from outside the derrick on to the derrick floor.

hog (mud): slush pump for circulating the mud in a well bore *(see Duplex pump).*

hole (crooked): well bore which is unintentionally deviated or 'corkscrewed'.

hole (dry): non-productive well also known as a 'duster'.

hook: heavy duty hook of 500 tons (508 tonnes) or more capacity which is suspended from the travelling block *(see travelling block)* and handles the drill pipe and casing loads when drilling or when 'running in' *(see run in)* or 'pulling out' *(see pull out).*

hook (wall): fishing tool used for pulling the top of a 'fish' *(see fish)* into the centre of the hole to permit the engagement of an 'overshot' *(see overshot)* or other fishing device to recover the 'fish'.

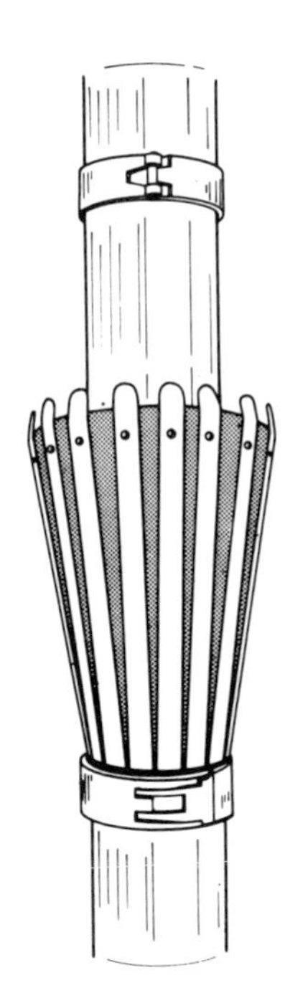

Cement Hopper

hopper (cement): funnel shaped container for dry cement fitted at its base with a nozzle through which water is injected at high pressure to form a cement slurry for the purpose of cementing a string of casing *(see casing string)* or setting a 'cement plug' *(see cement hopper).*

horizon (oil): interface between oil and gas or oil and water in a reservoir.

horse head: the fitting at the end of a 'walking beam' *(see walking beam)* of a pumping jack *(see pumping jack)* to which the sucker rods are attached by wires and an elevator to operate the down hole pump in a 'pumping well' *(see on the pump).*

hose (rotary): high pressure rubber or steel hose some 60 ft (18.3 m) long which connects the 'stand pipe' *(see stand pipe)* to the rotary swivel for conveying circulating mud to the drill string.

house (dog): drillers' office on the derrick floor which houses the 'lazy bench' *(see lazy bench)* and the 'knowledge box' *(see knowledge box)* or desk.

hydrocarbon: a compound made up only of molecules of hydrogen and carbon.

I

impression block: a tube filled with lead or wax which is run into the hole to obtain an impression of the top of a 'fish' *(see fish)* in order to assess the most suitable method of recovering the 'fish'.

inhibitor: an additive to reduce corrosion in the casing when pumping acid during an 'acid job' *(see acid treatment)* to increase the flow of hydrocarbons from a limestone formation. Chemicals used for this purpose may be salts of chromate and sodium nitrate added to the acid mix.

injection well: a well used for injecting gas, oil or water into the reservoir formation. When this exercise is required, special injection wells are drilled into the appropriate part of the reservoir. Gas injection serves to maintain the field pressure and provides a means of disposing of surplus gas which is separated from the oil *(see gas – natural)* produced from a well. Injection pressures may be very high (2 000 to 3 000 lbs/in^2 – 141 to 211 kg/cm^2 – or more) according to the field pressure which must be overcome and volumes may be large according to the amount of surplus gas available. Special compressors are used and the capacity of a typical installation is:

H.P. input: 30 000 to 110 000 H.P. (22 371 to 82 027 kW)
Injection pressure of gas: up to 9 000 lbs/in^2 (633 kg/cm^2)
Volume of gas handled: up to 500 000 000 cu.ft (14 000 000 m^3) per day.

Oil is sometimes injected into a suitable porous formation covered by a cap rock *(see cap rock)* and stored in large quantities for use at some future time. Gas may also be stored in this manner. When water is injected it is usually for the purpose of water flooding a field *(see water flooding)* to lengthen the productive life of the reservoir by flushing out oil which would not otherwise be recovered.

inner core barrel: the innermost barrel of a double or triple core barrel which retains the core cut from a formation *(see core).*

J

jacket: the lower section of an offshore platform which is fixed to the sea bed by piles and is mainly below water level.

jackup: an offshore drilling rig with a floating hull similar to a ship and fitted with retractable legs which can be lowered to the sea bed and elevate the hull structure above wave level. These units are suitable for water depths to around

350 ft (107 m) beyond which depths it is necessary to make use of 'floating rigs' *(see floating rig)* or 'semi-submersible' type rigs *(see semi-submersible rig).*

jar: a 'fishing tool' *(see fishing tools)* which when matched up to the top of a 'fish' *(see fish)* is used for striking heavy upward blows. Jars are also run above the bit in a 'cable tool string' *(see drilling – cable tool)* to provide a means of striking an upward blow to the bit which may have a tendency to jam in the hole being drilled. The jars used in rotary drilling *(see drilling – rotary)* are usually 'hydraulic jars' which are operated by an impact from fluid in one chamber of the jar flowing under high pressure through a restricted opening into another chamber. The jar is 'set' by lowering the drill pipe over the stroke length of the tool and then pulling up to cause the jar effect. Cable tool jars are simpler in construction consisting of two interlocking links which are pulled sharply upwards by the operating wire line to strike a blow.

joint (safety): *(see safety joint).*

K

kelly: a square or hexagonal pipe some 35 ft (10.7 m) long which is screwed to the top of the drill string and itself is supported by the 'swivel' *(see swivel)* suspended from the main hook and travelling block and is capable of being raised or lowered over its entire length whilst rotary motion is transmitted to the kelly by the 'rotary table' *(see rotary table).*

kelly bushing: a driving assembly which transmits rotary motion to the 'kelly' *(see kelly)* from the rotary table and also permits the kelly to be raised or lowered over its entire length when rotating or stationary.

kelly cock: a cock placed between the top of the kelly *(see kelly)* and the swivel *(see swivel)* to provide a means of closing off any flow of fluid from the well through the drilling string in an emergency situation.

key seat: a groove in the well wall which may cause difficulty in 'pulling out' of the hole due to the bit or tools on the bottom of the drill string holding up in the 'key seat'.

killing a well: the operation of controlling a wild well or 'mudding off' *(see mud off)* a completed well to hold the reservoir pressure in check.

knowledge box: the desk in the 'dog house' *(see dog house)* or drillers' office on the rig floor.

knuckle joint: a directional drilling tool run in the drill string and controlled by hydraulic pressure of the circulating mud to set the bit or tools at an angle in the hole.

L

land (casing): the situation when a casing string *(see casing string)* is hung or flanged in its final position.

landing joint: the top joint of a casing string *(see casing string)* which extends from the point at which the top of a string is to be set and up through the rotary table and accommodates the cement head *(see cement head)*. After the cement slurry has been pumped into the well and has set, the landing joint is removed leaving the cemented string of casing in the hole.

latch on: to attach the elevators *(see elevator)* to the drill string.

lay barge: special barge equipped to handle the laying of submarine pipelines *(see pipeline – submarine)*.

lazy bench: long wooden seat in the 'dog house' (or drillers' office on the derrick floor level).

lead tong: a pipe wrench which hangs in the derrick and is operated by a wire line from the 'cathead' *(see cathead)* to make up or break out joints of pipe from the drilling string.

liner: a string of casing which is run into the production area of a well to protect the face of the formation and prevent sand or debris from flowing into the well. A liner is usually hung using a liner hanger *(see liner hanger)* set in the lower section of the last casing string in the hole.

liner completion: where a reservoir formation is somewhat unconsolidated a perforated length of casing may be hung by a liner hanger in the lower joint of the final casing string in the well. The liner extends through the producing formation and prevents the intrusion of particles of the rock formation as the oil flows into the well. Such liners may be slotted before they are run into the hole or they may be perforated by shooting bullets from a special gun after the liner has been landed in position.

liner hanger: assembly fitted with 'slips' *(see slips)* and 'packers' *(see packer)* which is used for the hanging of a 'liner' *(see liner)* in the lower section of the production casing string.

live oil: oil that contains gas.

log (drilling): the record of the drilling operations recorded by the driller in each 'tour' *(see tour).*

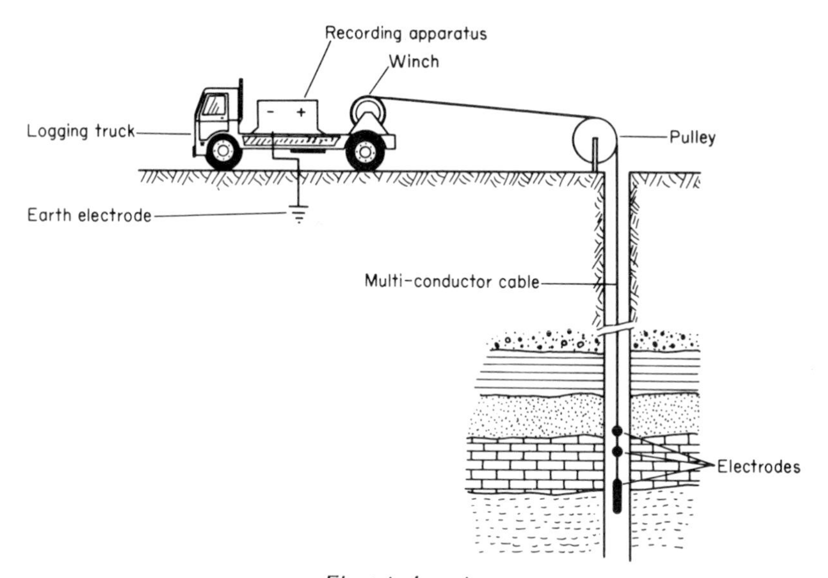

Electric Logging

log (electric): by using electric logging a tremendous amount of information is obtained from a completed well giving details of the types and location of formations penetrated, the porosity, the permeability, the 'dip' *(see dip meter),* type of fluid contained in a formation etc. Electric logs are extremely accurate in depth recording to within a few inches and the information obtained will influence decisions on production procedure and casing programmes. The information is obtained by running special tools into the well on an electric cable and recording on film at surface the effect of passing an electric current through the tools to record the differential resistance of various formations.

log (well): record of geological formations penetrated during drilling and including technical details of the operation.

logging (mud): this exercise provides a continuous examination of the mud which is being circulated in a well and provides evidence of oil or gas which may be present in formations being drilled. Instruments are usually installed in a trailer type laboratory alongside the well and a continuous record is made giving pump delivery details, mud weight figures, oil or gas content in the mud and an analysis of the mud properties.

lose returns: a situation where the circulating fluid enters a porous formation or 'cave' and does not return to the surface in the normal way.

lubricator: a special cylindrical container fitted with valves which facilitate the running of wire line tools into a well under pressure or the injection of mud or other fluids into the well under similar pressure conditions.

M

macaroni: small diameter drill pipe of 2⅜ inch or 2⅞ inch (60 mm or 73 mm) outside diameter.

magnet (fishing): a permanent magnet run on the end of a drilling string to recover small steel junk which may be present in the hole.

make it up another wrinkle: to tighten a connection one more turn.

making a connection: the operation of adding a single joint of drill pipe to the drilling string after the 'kelly' *(see kelly)* has been drilled down its full length.

making a trip: the operation of pulling the drill pipe out of the hole, changing the bit or tools and re-running the pipe to the bottom. In deep holes this can take 12 hours or more.

marsh funnel: a funnel shaped instrument used for determining the viscosity of the circulating fluid.

The funnel is 6 inches (152 mm) in diameter at the top and 12 inches (305 mm) long and a 10 mesh screen is fitted over half of the top to remove cuttings from the mud sample as it is poured in. The funnel is filled whilst holding a finger over the outlet tube at the bottom. The finger is then removed and the time in seconds for a volume of one quart (1.13 litres) of fluid to flow out is recorded. Results obtained from the Marsh Funnel cannot be correlated directly to that of a rotary viscometer but the instrument is very suitable for rugged use on a rig *(see rig – drilling)* and is easy to use.

Using a Marsh Funnel fresh water has a viscosity of 26 seconds and many good muds have viscosity values between 34 and about 50 seconds.

mast: the steel structure erected over a well to handle the lifting and lowering operations of a drill string or casing string *(see casing string)* and is usually designed for easy raising or lowering into position and can be conveniently transported from location to location more easily than is the case with a 'standard derrick' *(see derrick)* structure which must be assembled in sections.

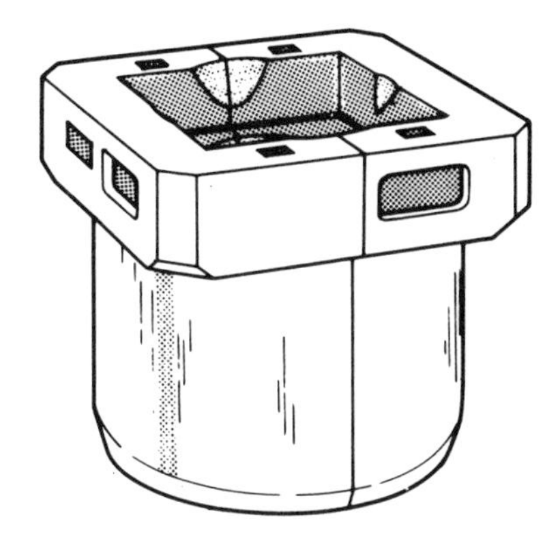

Master Bushing

master bushings: heavy steel inserts which fit into the rotary table *(see rotary table)* to support the weight of the drill string by means of removable tapered slips *(see slips)*. The master bushings are easily removable from the centre of the table to permit large diameter tools or fittings to pass through the table aperture.

master gate: the main valve fitted to a wellhead of a production well.

measure in: measuring the drill pipe with a steel tape when running into the hole to check the true depth of the hole.

measure out: measuring the drill pipe with a steel tape when pulling out of the hole to check the true depth of the hole.

module: large containers in which are housed the various units of equipment such as power packs, pumping sets, control equipment, sewerage plant etc for installation on a production platform *(see production platform)*. The plant or equipment is assembled and fitted into the modules on shore to be transported as units for installation on the platform. These units may in some cases weigh as much as 1 800 tons (1 830 tonnes).

monkey board: the platform used by the derrick man *(see drilling crew)* some 90 ft (27.4 m) above the derrick floor to handle stacking of the stands *(see stand)* when 'pulling out' or 'running in' the drill string.

Mother Hubbard: a type of cable tool *(see drilling – cable tool)* drilling bit.

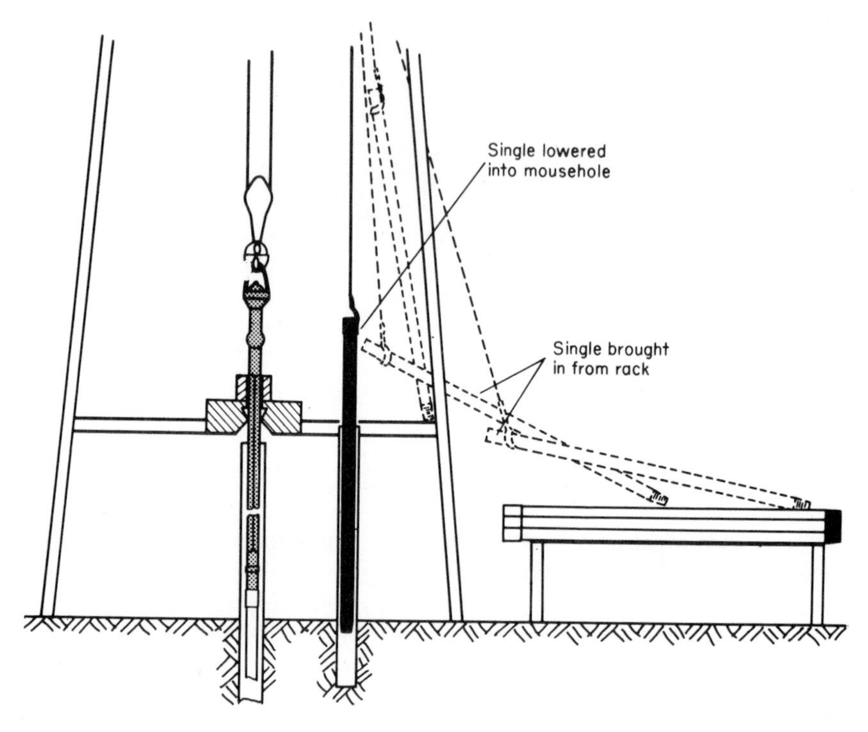

Mouse Hole

mouse hole: a shallow hole (about 25 ft (7.6 m) deep) drilled near the rotary table *(see rotary table)* to accommodate a single joint of drill pipe for the purpose of speeding up the operation of adding single joints of drill pipe to the drilling string without the time consuming operation of setting back the kelly *(see kelly)* into the rathole *(see rathole)* when making a connection *(see making a connection).*

mud: fluid used for circulating a well during a drilling operation and which is carefully controlled in its properties by the addition of chemicals and additives to provide correct properties of specific gravity, viscosity, fluid loss *(see mud – cake)* gelling capability *(see gel)* etc to facilitate the drilling and control of pressure conditions existing or anticipated. There are certain basic constituents used in making a standard 'mud' but the type of formation being drilled and pressure or other conditions in any particular well will dictate the control of the mud properties by the addition of one or many chemicals.

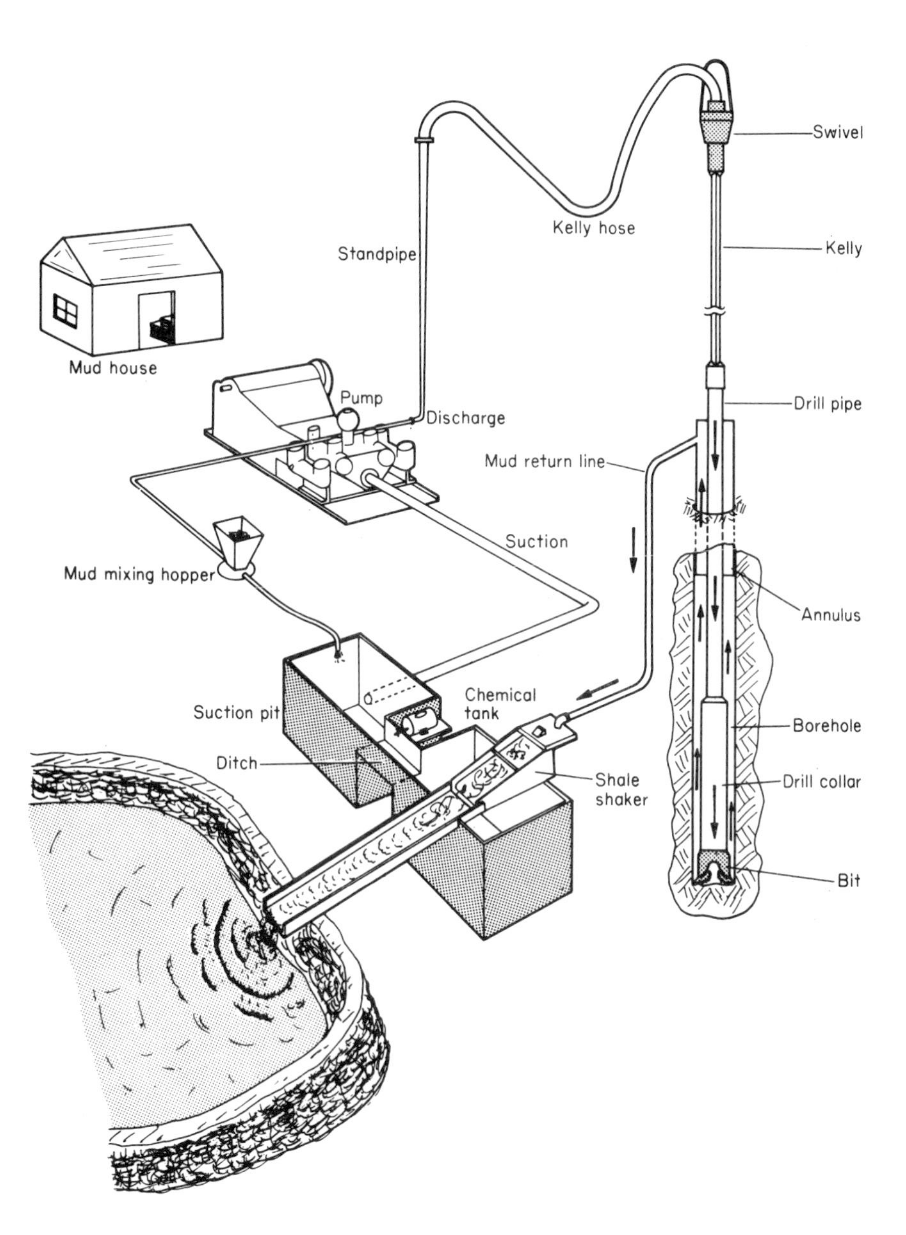
Swivel
Kelly hose
Kelly
Standpipe
Mud house
Pump
Discharge
Drill pipe
Mud return line
Suction
Mud mixing hopper
Annulus
Chemical
tank
Suction pit
Borehole
Ditch
Shale
shaker
Drill collar
Bit

Mud System

Mud is required for the following purposes:
(a) to cool the bit;
(b) to remove cuttings from the hole;
(c) to prevent the hole from caving;
(d) to consolidate loose formations such as gravel or sand;
(e) to prevent the intrusion of water, gas, or oil into the hole;
(f) to lubricate the drill pipe;
(g) to check corrosion of the drill pipe and casing;
(h) to suspend cuttings in the hole when drilling is stopped for any reason. The gel property *(see gel)* of a mud column prevents cuttings from settling to the bottom of the hole and causing the bit and drill collars to become stuck when circulating is suspended;
(i) to permit cuttings and sand to be removed by the vibrating screen *(see vibrating screen)* and settling pits before re-pumping the cleansed fluid to the well.

Constituents: fresh water is the most usual base but saturated salt water is used when drilling through salt formations in order to prevent the absorption of salt from the formation which would destroy the mud properties and cause 'leeching' out of the well walls.

In some special cases an 'oil base mud' is used or an 'oil emulsion mud' but these muds are expensive and 'dirty' to handle and are not popular unless they are essential. Their use is rare in most areas.

A colloidal clay such as bentonite (silicate of Ca, Mg and Al with H_2O) is added to the water base to form a fluid which has good gel forming, thixotropic, non-corrosive, non-abrasive and lubricating properties. Bentonite alone will not, however, make a mud which is heavier than about 8.5 lbs per gallon (823 g per litre) and must be 'weighted' with other materials if the mud column is required to control gas, oil or water pressures encountered during drilling.

The most common material used for 'weighting' a mud is ground barite ($Ba\ SO_4$) which is added to the fluid by means of a hopper *(see hopper – cement)*. Below are listed some of the chemicals which may be added to the circulating fluid for various reasons:

(a) sodium hydroxide (caustic soda Na_2OH) maintains the pH value and prevents the growth of micro-organisms in a starch mud.
(b) sodium carbonate (soda ash Na_2CO_2) increases gelling properties or controls the pH value and is used to treat anhydrite or gypsum contaminated fluids.
(c) sodium bicarbonate (baking soda) ($Na\ HCO_3$) counteracts the effect of cement contamination.
(d) sodium silicate (water glass) ($Na_2\ SC_4\ O_9$) used for handling a sloughing shale *(see shaker – shale)*.
(e) calcium hydroxide (lime) $Ca\ (OH_2)$ used in 'lime muds' which are less

affected in viscosity when drilling shales or salt water flows than are phosphate treated fluids.

(f) barium carbonate ($BaCO_3$) used to precipitate anhydrite or gypsum.

(g) sodium acid pyrophosphate ($Na_2H_2P_2O_7$) used to treat cement contaminated mud.

(h) tetra sodium pyrophosphate ($Na_4P_2O_2$) used as a 'thinner'.

(i) tannin ($C_{14}H_{10}O_9$) extensively used as a mud thinner.

It must be appreciated that 'mud engineering' is a very specialised field and the circulating fluid is continuously analysed whilst drilling a well and suitable adjustments made by experimentation in a laboratory before full scale treatment is introduced. In some cases it becomes necessary to suspend drilling operations for the purpose of treating the mud in circulation or even replacing it with an entirely new type of mud. Either of these operations may cost thousands of pounds in materials and lost 'rig time'.

mud (cake): a deposit of solids from a drilling mud which builds up on a well wall opposite a porous formation due to the loss of water from the mud to the formation. The porous formation acts in effect as a filter element. This condition can cause serious complications in that it may become impossible tc pull the bit out of the hole if the mud cake builds up sufficiently to obstruct the hole.

mud (gas cut): mud fluid contains entrained gas whereby the apparent specific gravity of a given volume of mud is reduced and the hydrostatic pressure of a column of the mud will be below that of a similar column of normal mud. This is a condition which could become dangerous and allow the well to 'blow out'. Small quantities of gas can be removed by 'gunning' the mud *(see gun – mud)* before returning it to the well or in extreme cases it may be necessary to pump a new volume of mud and thus to remove the gas cut mud in circulation.

mud box: the tank or container from which the circulating pump sucks mud to deliver it to the well.

mud ditch: a ditch or fluming through which mud returns from the vibratir screen to the circulating pump suction tank.

mud gun: *(see gun – mud).*

mud hog: *(see Duplex pump).*

mud logging: *(see logging – mud).*

mud mixing plant: a combination of pumps, piping manifold, 'hoppers' *(see hopper – cement)*, guns *(see gun – mud)* and tanks which is necessary for mixing the constituents making up a drilling mud.

mud off: the procedure used to 'kill' a well or prevent the formation fluids from entering the well bore.

mud pit: the pit or tank into which return fluid from the well discharges. Also mud pits are used for storing reserve volumes of mud.

N

natural gas: *(see gas – natural).*

O

offset: a well bore which is set off from the vertical *(see directional drilling).*

offshore structure: steel or concrete structure placed on the sea bed and installed for the purpose of supporting a drilling rig or oil or gas production equipment.

oil field: usually defined as an area where productive oil deposits have been proved.

oil horizon: *(see horizon – oil).*

oil sand: a porous sandstone formation containing oil.

oil show: indication of an oil reservoir by the presence of oil in the cuttings returning to surface or oil flowing into the well bore.

oil string: the casing string *(see casing string)* which is run to keep the well open and allow the production of fluid from an oil bearing formation to surface.

on the pump: a well which will not flow on account of a low reservoir pressure and which must be fitted with a down hole pump.

open hole: the uncased portion of a well.

out step well: a well drilled to explore the extent of a field beyond the proved area.

overshot: a 'fishing tool' *(see fishing tools)* usually fitted with 'slips' *(see slips)* having radial teeth and which is run over the top of a string of pipe lost in a hole to latch on and recover the fish.

P

packer: an expanding plug, usually made of rubber which is run and set in a hole to obstruct the passage of fluid.

pay sand: a sand reservoir which contains oil or gas of productive value.

percussion drilling: *(see drilling – percussion).*

perforating: piercing the casing by use of a gun *(see gun – perforating)* opposite a producing formation to allow the oil to flow into the casing for production of the well.

permeability: the property of a formation which allows or obstructs the flow of a fluid contained to pass through the pore spaces of the rock and into the well bore. A rock may be very porous but may have poor permeability, i.e. the microscopic passages between the pores may be restrictive to flow.

persuader: a large tool used to exert excessive pull when unscrewing a very tight joint of pipe or some such situation.

pig: a device or 'go devil' used for pumping through a pipeline to clean the walls of the pipe or clean an obstruction *(see rabbit).*

pilot bit: bit used for starting a directional hole, i.e. a hole deviated from the true vertical *(see directional drilling).*

pin: male thread tapered connection.

pipe rack: a staging outside the derrick floor on which drill pipe or casing is stored and can be easily pulled into the derrick floor via the 'pipe ramp' *(see casing ramp).*

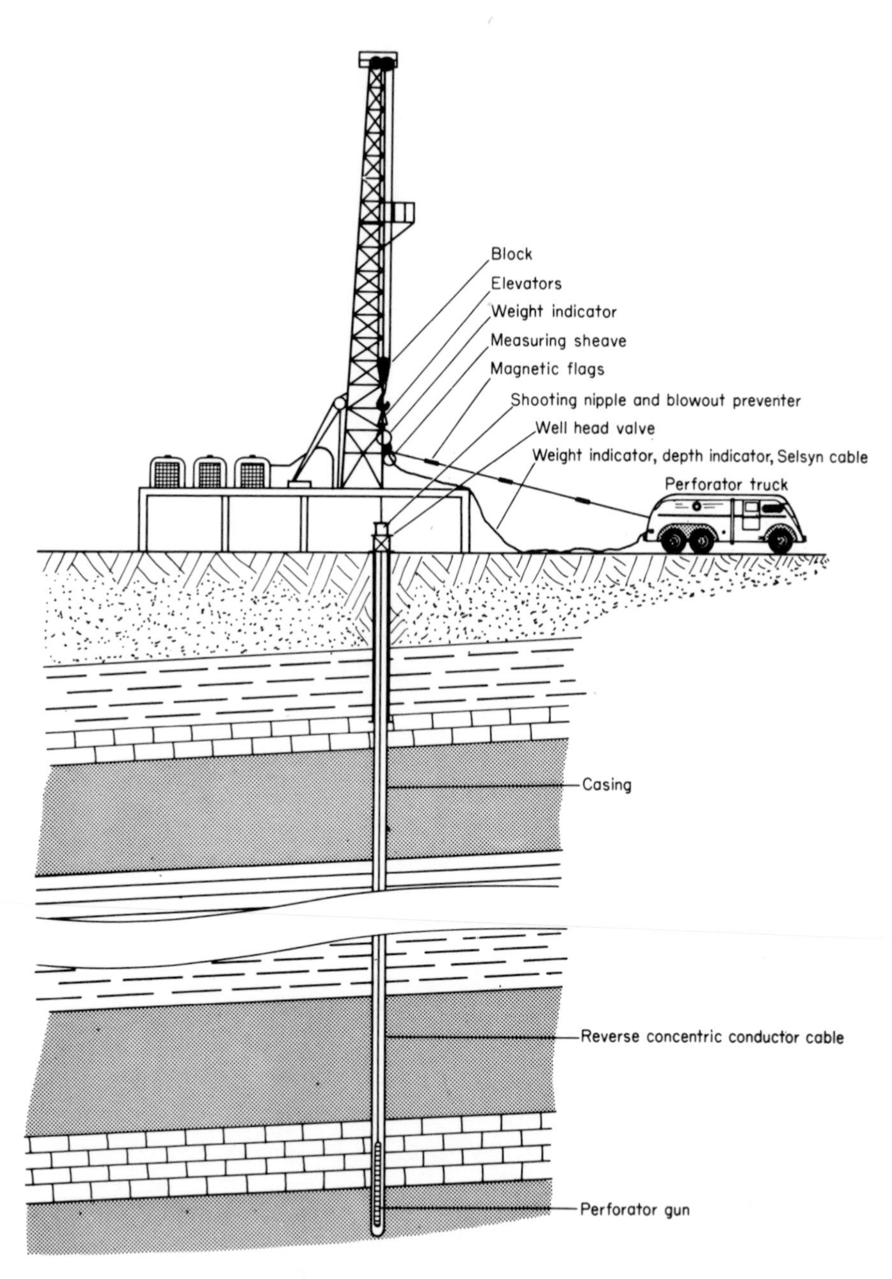

Layout of equipment to perforate a well

pipeline (submarine): the most satisfactory way of transporting oil or gas from an offshore location is by means of a large diameter pipeline (32 ins to 34 ins (813 mm to 864 mm) diameter). Such pipelines are 'trenched into' the sea bed and are laid by specially constructed 'lay barges', on which the pipe lengths are welded together and fed off via a 'stringer' or ramp at the stern of the barge. Before loading on to the lay barge the pipe lengths are coated with a weighted cement aggregate of some 2 inches (51 mm) thickness in order to protect the pipe and provide weight to overcome its buoyancy.

Such pipelines have never previously been laid in depths of water existing in the North Sea and the entire operation is an extremely expensive exercise often running into figures of £1,000,000 per mile laid (at 1975 prices).

pipeline oil: oil which is clean enough to be marketed.

pirate sand: a porous sandstone which allows fluid to escape from the bore hole.

pitman: the rod which connects the 'walking beam' *(see walking beam)* to the drive crank on a percussion drilling rig *(see drilling – percussion).*

plug back: to place a cement plug at the bottom of a hole to seal off an area for one of any number of reasons.

polished rod: a highly polished rod which reciprocates through a gland on the well head to operate a down hole pump or down hole tools in a well which may be under pressure condition.

poor boy: a tool or operation which is done 'cheaply' instead of making use of the most sophisticated equipment or service available.

porosity: the percentage of void in a porous rock compared to the solid formation.

potential test: test which provides information on the productive capacity of a well.

production platform: once the exploration drilling programme has been completed and when a field *(see field)* has been established, a production platform (or a number of platforms) are positioned over the field to handle the drilling of the production wells and the production of oil or gas from the reservoir. Such platforms, either of steel or concrete construction, are extremely

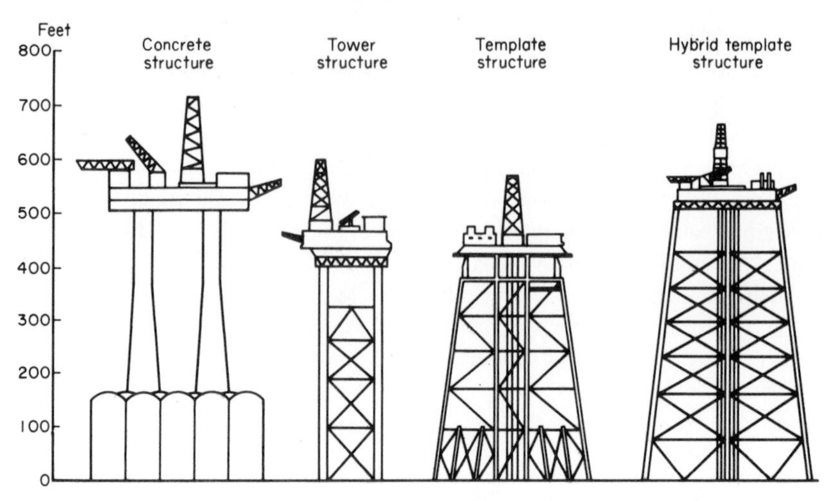

Types of offshore production platforms

expensive to construct and position. At present day prices a single unit, complete with the necessary equipment for handling the oil, may amount to some £68,000,000 and one such platform will serve approximately an area of a two mile diameter circle of the oil bearing formation by the drilling of 27 to 30 directional wells *(see directional drilling).*

Platforms are constructed in deep water bases as near the field as possible and towed to site and accurately positioning on the sea bed is a major operation particularly when the North Sea weather conditions must be taken into account. Steel structures, once located, must be secured by piles which may penetrate into the sea floor by as much as 250 ft to 300 ft (76 m to 91 m). Concrete structures need a firm level bed on the sea floor and extensive surveys of the soil conditions must be carried out before a decision can be made as to the suitability of a site.

The cost of constructing a platform and setting it in position increases tremendously as water depths increase and under present conditions and technical 'know-how' it is believed that such units are neither economical nor possible to construct for water depths much greater than 600 ft (183 m).

When in position and permanently installed, two drilling rigs are set up on the deck which may be some 110 ft (33.5 m) above sea level.

Directional wells are then drilled and some of these will bottom at points a mile or so from the site of the platform thus serving to take production from a two mile circle of the oil bearing formation. Ideally a well will take about six weeks to drill and complete and as two rigs are running at the same time one well should be completed every three weeks. This situation presents tremendous problems in connection with keeping supplies of mud chemicals, cement, and casing to the rigs especially as weather conditions greatly restrict the period when supplies can be shipped and offloaded on to the platforms. Any delays can be very costly as a drilling rig costs £15 a minute or more to operate and, consequently, shut down periods are to be avoided at all costs.

Whilst the wells are being drilled and completed the modules *(see module)* containing the power units and production equipment for the production of the oil or gas are loaded on to the deck and installation proceeds.

On completion of the drilling programme there will be 27 to 30 wellhead Christmas trees *(see Christmas tree)* located immediately below the deck of the platform. Production from these wells will be flowed through separators to remove the gas entrained in the oil, and the oil and possibly the gas will then be pumped into pipelines for delivery to shore bases.

In some cases excess gas must be re-injected via specially drilled wells into the reservoir formation in order to dispose of surplus gas and at the same time to assist in maintaining the field pressure and so lengthen the life of the field. In some cases water will be pumped into the field through specially drilled wells in order to provide a 'water drive' *(see water drive)* and so increase the final recovery of reserves.

All this equipment requires a tremendous amount of power to operate the high pressure pumps and compressors and this power will be supplied most probably by gas turbines running on gas from the oil production. A single platform may have a power installation which is sufficient for the needs of a small town. *(See Appendix)*

Previous to North Sea discoveries, production platforms had never been constructed and installed in such deep water, and rough weather conditions and the hazards to which they will be subjected give rise to serious concern. Many engineers believe that other less hazardous methods of production must be designed and a number of companies and organisations are pressing for perfecting 'sub-sea completions' *(see sub-sea completion)* whereby the wellhead Christmas trees are on the sea floor and the danger of a disaster situation with a platform would not cause damage to the wells themselves and the possibility of a serious 'blow out' *(see blow out).*

In this connection it is interesting to note that the US Geological Survey in a notice in the Federal Register asks industry to assess the status of technology in sub-sea well completion techniques and production systems.

The Department of Interior agency also wants forecasts of development in the field and is interested in learning a reasonable time frame for the application of sub-sea technology.

propping agent: material such as walnut shells, glass balls or coarse sand which is injected into a formation to keep a fracture open when a reservoir formation is broken down by very high pressure pumping to increase the permeability factor *(see fracturing – formation).*

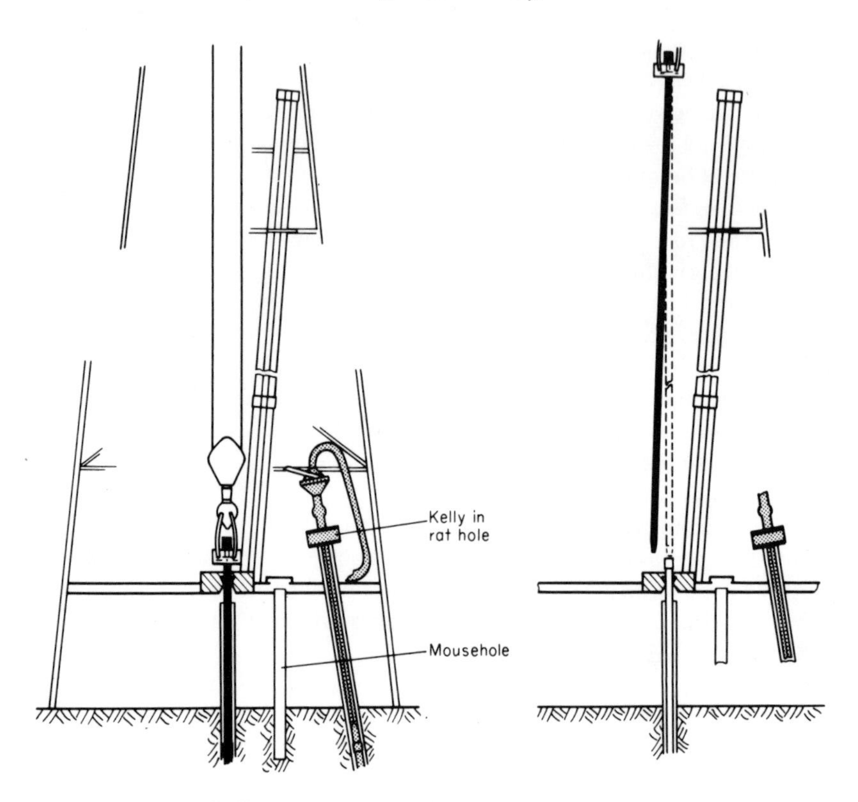

Pull out. Hoisting drilling string out of hole

pull out: to hoist the drilling string out of the hole.

pump off: the situation where a pumping well *(see put on pump)* is pumped to a degree where the oil level falls below the level of the pump standing valve.

pumping jack: the operating unit erected over a pumping well *(see put on pump)* which operates the down hole pump *(see down hole pump).*

put a well on: to open up a well to production either by pumping or flowing into the collecting system.

put on pump: to install pumping equipment in a well which will not flow to surface under the formation pressure.

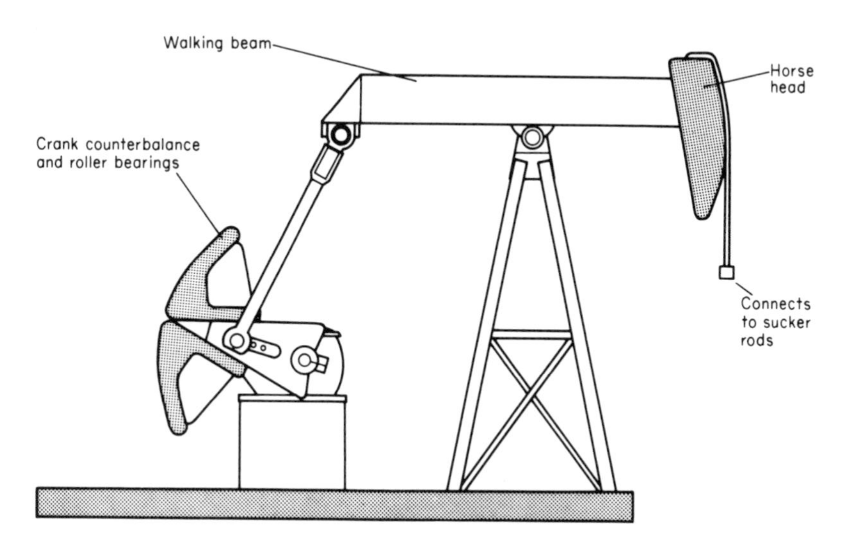

Pumping Jack

Q

quebracho: thinning agent added to the mud to decrease its viscosity, otherwise known as tannic acid ($C_{14}H_{10}O_9$) which is produced from the bark of a tree.

R

rabbit: similar to a 'pig' *(see pig)* which is pumped through a pipeline to clean or test for obstructions.

rathole: a shallow hole drilled near the corner of the derrick floor which is used to accommodate the kelly *(see kelly)* when it is removed from the drill string and is not in use.

ream: to enlarge a hole which is under size for running casing or down hole tools.

reamer: a tool fitted with serrated rollers or expanding cutters which is used for enlarging the diameter of a bore hole.

reservoir: a porous formation containing fluid or gas and which is sealed by a 'cap rock' *(see cap rock)*.

reverse circulation: a procedure adopted in special circumstances when the circulating fluid is pumped down the annular space between the well wall and the drill pipe and is returned through the drill string.

Reverse circulation may be used when a large diameter hole is being drilled and the pumping capacity for the circulating fluid is insufficient to give a rising velocity of fluid in the annular space to float the cuttings to surface (the ideal rising velocity should be around 100 ft (30.5 m) per minute). In such cases the mud may be pumped into the annular space between the drill pipe and the well wall and the return through the drill pipe will provide a much higher rising velocity due to the small diameter of the bore of the drill pipe. In these circumstances a rotary packer must be fitted to the wellhead to seal the annular space and allow the drill pipe to be rotated for the drilling operation.

rig (drilling): the term 'drilling rig' covers the main drilling installation consisting of a derrick or mast *(see derrick)* substructure, drawworks *(see drawworks)*, engines, pumps, boilers (if any), rotary table *(see rotary table)*, crown block *(see crown block)*, travelling block *(see travelling block)*, main

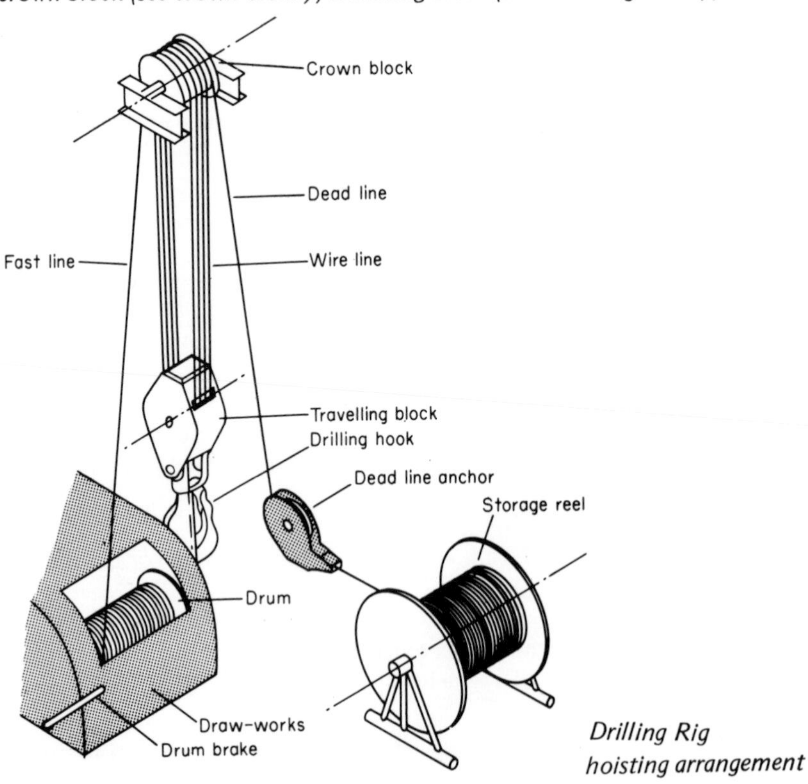

Drilling Rig hoisting arrangement

hook *(see hook)*, vibrating screen *(see vibrating screen)*, storage tanks, cementing equipment, mud mixing plant, and all tools required for handling the drilling of a well.

NOTE. The term 'rig' is frequently used by the 'layman' and misinformed reporters when referring to what is, in fact, a 'production platform'.

rig up: installing the rig *(see rig – drilling)* and necessary equipment for drilling a well.

riser: a pipe through which fluid travels in an upward direction. On an offshore operation the term 'riser pipe' refers to the large diameter pipe (16 in (406 mm) or so) which extends from the blow out preventer stack *(see blow out preventer stack)* on the sea floor to under the derrick floor of a semi-submersible rig *(see semi-submersible rig)* or a floater *(see floating rig)*.

rock bit: the cone type bit which is run at the lower end of a drill string to attack the formation being drilled.

rock hound: a geologist.

rock pressure: the pressure of the fluid in a reservoir.

rocking: the operation of pressurising and releasing pressure in cycles to induce a well to flow.

roller bit: cone type bit *(see rock bit)*.

rope spear: fishing tool *(see fishing tools)* with projecting barbs used to recover a broken wire line from a hole.

rotary bushing: heavy steel liners which fit into the rotary table *(see rotary table)* opening to accommodate the 'kelly bushing' *(see kelly bushing)* or the 'pipe slips' *(see slips)*.

rotary drilling: *(see drilling – rotary)*.

rotary hose: *(see hose – rotary)*.

rotary table: the rotary table is situated in the centre of the derrick floor immediately over the well bore and directly under the centre of the derrick or mast. Its function is to support the weight of any pipe or casing run into the hole and to provide rotary motion to the drillpipe string via the kelly *(see* *kelly)* when drilling. The table which resembles a solid fly wheel mounted horizontally and having teeth on its underside is rotated by a pinion and chain or shaft drive from the drawworks *(see drawworks)*.

The centre of the table has an opening into which 'master bushings' *(see master bushings)* fit to provide a recess for the 'kelly bushings' *(see kelly bushings)* or to accommodate the drill pipe 'slips' *(see slips)* which support the drilling string when it is being run into or pulled out of the hole.

rough neck: *(see drilling crew).*

round trip: the operation of pulling the drill pipe out of the hole, changing the bit or tools and re-running the pipe to bottom. The time taken for a 'round trip' will depend upon the depth of the hole and the conditions prevailing in the hole, such as caving formations or pressure shows existing in water, gas or oil reservoir rocks.

In a deep hole (say 15 000 ft – 4 572 m) a round trip may take some twelve hours or more.

roustabout: *(see drilling crew).*

run in: to run drill pipe, casing or tubing into the hole.

running sand: grains of sand suspended in water or oil which can cause complications when drilling a hole.

S

safety joint: a safety tool with a fast left hand thread which is run in the drill pipe string above the bit or a 'fishing tool' *(see fishing tools)* and which permits the drill string above the safety joint to be easily backed off and recovered in the event of the bottom hole tools becoming stuck in the hole.

samples: cuttings from the formation being drilled which are returned to surface in the circulating fluid and removed on the vibrating screen *(see vibrating screen)* for examination by the geologist.

Sampson post: upright post which supports a walking beam *(see walking beam).*

sand line: a wire line used to run a 'sand pump' *(see sand pump)* or bailer *(see bailer)* or a 'swab' *(see swab)* into the hole.

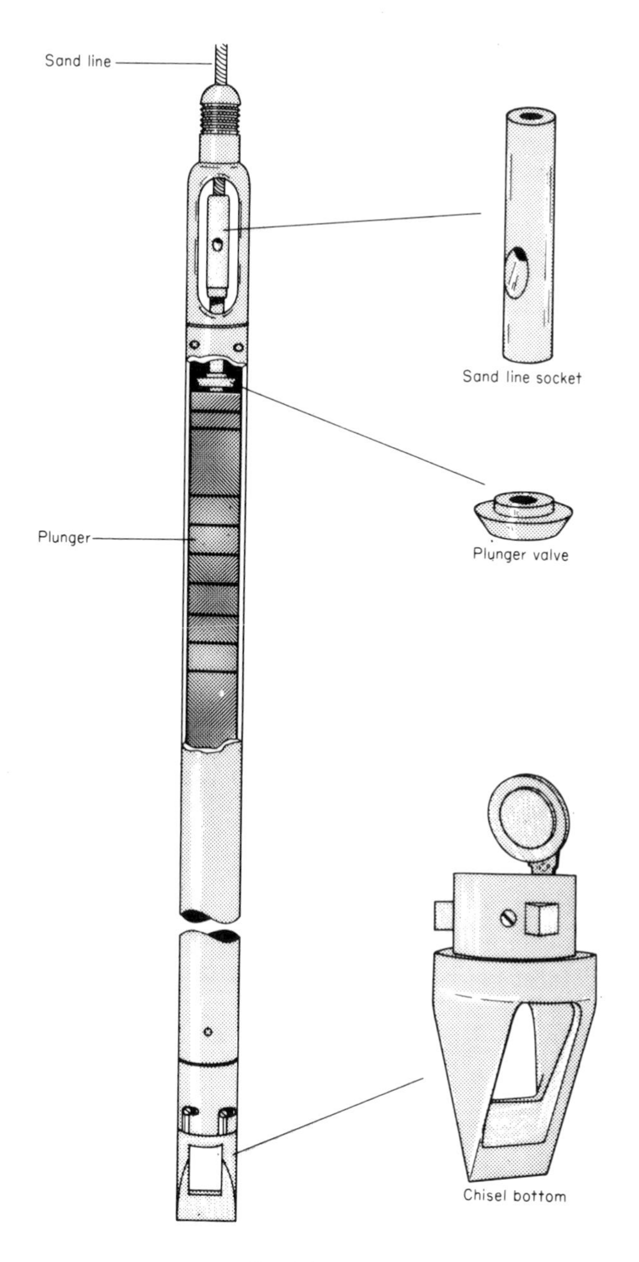

Sand Pump

sand pump: a type of bailer fitted with a plunger which is used for removing sludge and cuttings from a cable tool hole *(see drilling – cable tool).*

sand reel: the drum reel for handling the sand line *(see sand line).*

sand up: the condition where sand enters the well with the oil being produced and chokes the tail pipe *(see tail pipe)* or down hole pump *(see down hole pump).*

saversub: a short substitute which is run on the lower end of the kelly *(see kelly)* to connect with the drill string and which can be easily replaced when worn, thus protecting the bottom connection of the kelly.

Schlumberger: a household word in the industry to describe the means of electric logging of a bore hole (as the French scientist of that name was the original inventor of the system).

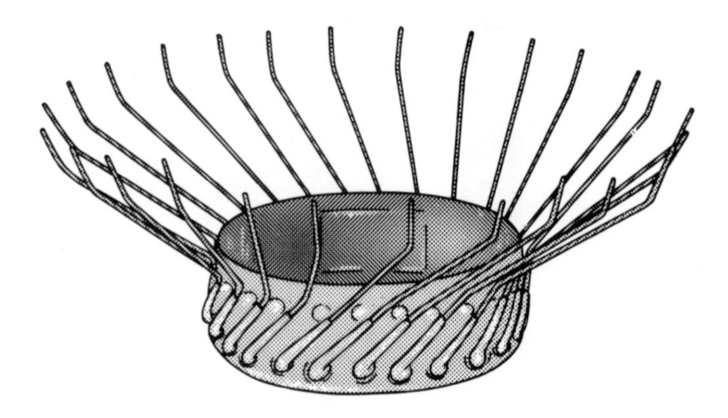

Scratcher

scratcher: a device fitted to a casing joint which cleans the well wall as the casing is being run and improves the hole conditions for a satisfactory cement job.

screen (vibrating): the 'shale shaker' used for removing cuttings from the returning circulating fluid and which allows the uncontaminated mud to return to the pump suction tank for recirculation in the well.

One or two wire mesh screens vibrate at high speed (driven by an electric motor) causing the cuttings to remain on the top of the screen and the fluid to flow through the mesh.

screen pipe: a short piece of perforated or slotted casing used to protect the face of a producing formation and to prevent sand or small pieces of the formation from flowing into the production string.

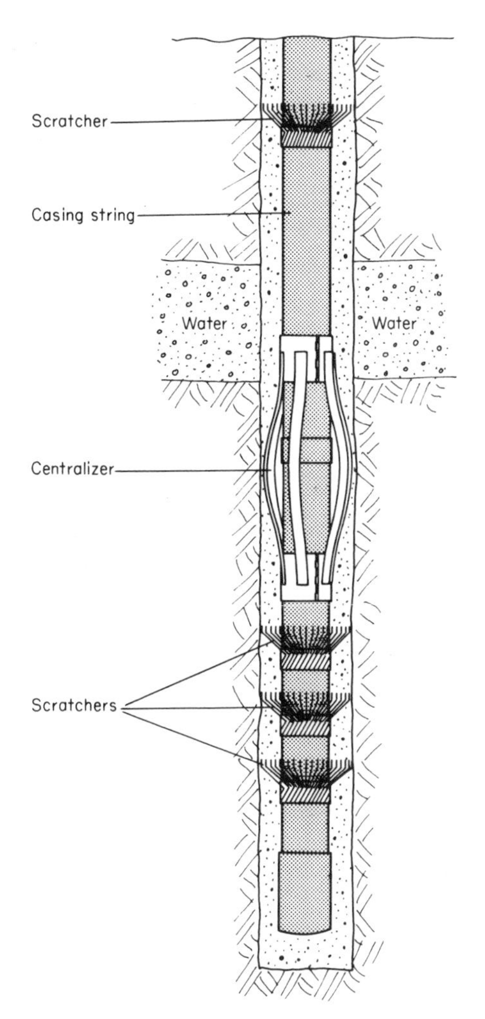

Array of Scratchers and Centraliser installed on the bottom joints of casing

secondary recovery: artificial recovery of oil by some means such as water injection *(see water flooding)* or gas injection. Normally water injection or gas injection *(see gas drive)* is instituted when a field is depleted to the stage where the production of oil due to the reservoir pressure is falling off as the pressure becomes expended and the well will no longer flow of its own accord. Water is pumped into wells drilled on the boundaries of the field and flushes oil from the reservoir formation into the well bore thus producing oil which would otherwise remain in the formation. Water flooding or pumping will, however, never provide a means of recovering all the oil which is present in a reservoir and, in fact, it is rarely possible to produce more than 40% of the oil in situ. The final recovery volume is dependent on the porosity and permeability factors of the reservoir formation.

seepage: a natural oil spring which occurs at surface where there is a fissure connecting to the oil reservoir. The presence of a seepage indicates that a reservoir formation exists in the vicinity but does not guarantee that a well drilled nearby will produce oil. It may be that the seepage has, over the ages, leaked most of the recoverable oil to the atmosphere.

seismograph: a device for detecting vibrations set up by creating an artificia earthquake in order to investigate the underground structure of the earth's formations. The source of the vibrations may be by charges fired in special shallow holes (say 100 to 200 ft (30.5 to 61 m) deep) or by using a 'thumper' i.e. dropping a heavy weight on the surface or in water by firing small charges from a cable towed behind a boat or by injecting high pressure air pulses into the water.

semi-submersible rig: an offshore drilling platform having large cylindrical shaped bodies located below wave level, thereby minimising the motion of the platform due to sea conditions at surface. Semi-submersibles are capable of drilling in very deep water and withstanding heavy storm conditions. They may be held on site by a computer which controls 'thrusters' located in the underwater cylindrical units or they may be anchored.

separator: vessel or device for separating mixtures of oil/water/gas.

set casing: the operation of landing a casing string and cementing it in place

shaker (shale): *(see screen — vibrating).*

shoe (casing): *(see casing shoe).*

shoe (drive): a short length of heavy wall pipe run on the lower end of a casing string *(see casing string)* which is to be forced into a hole.

shoe (float): casing shoe fitted with a non-return valve to facilitate floating a string of casing into a well and thus relieving the hoisting gear of excessive loads. A float shoe also provides the means of circulating mud through the casing string but will not permit a flow of fluid from a down hole formation upwards through the casing string.

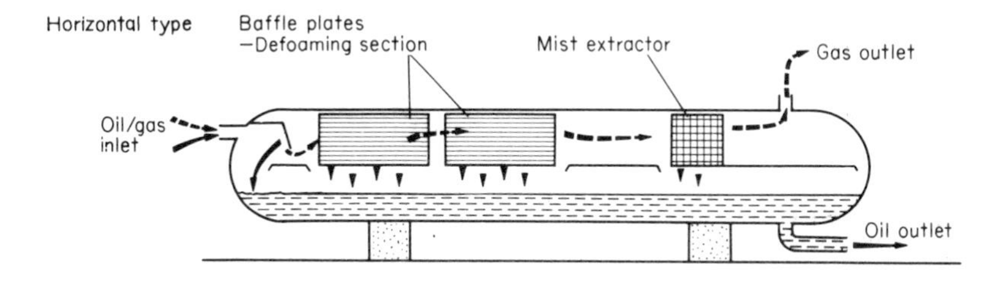
Horizontal type
Baffle plates
—Defoaming section
Mist extractor
Gas outlet
Oil/gas
inlet
Oil outlet

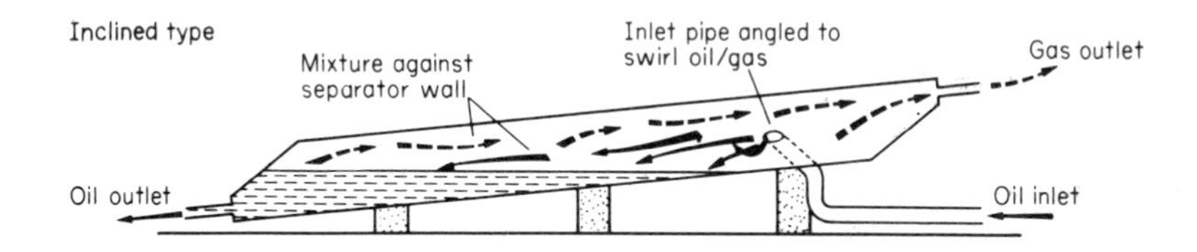
Inclined type
Mixture against
separator wall
Inlet pipe angled to
swirl oil/gas
Gas outlet
Oil outlet
Oil inlet

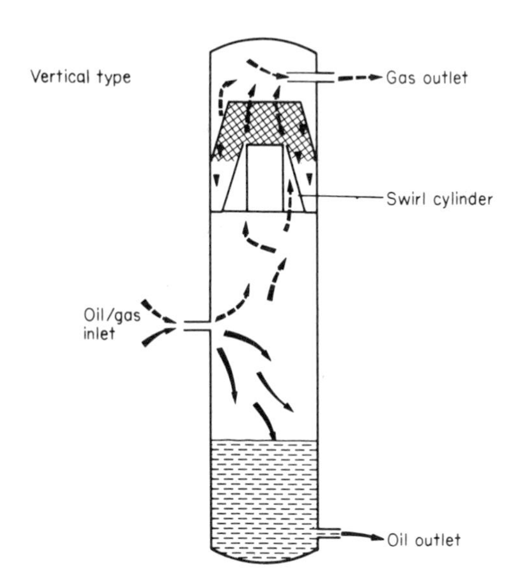
Vertical type
Gas outlet
Swirl cylinder
Oil/gas
inlet
Oil outlet

Separators

sidetrack: drilling past a 'fish' *(see fish)* which is permanently left in the hole for some reason.

side wall coring: *(see coring – side wall).*

single: one joint of drill pipe about 20 ft to 30 ft (6.1 m to 9.1 m) in length and fitted with a 'tool joint' *(see tool joint)* box connection on the top and a tool joint pin connection on the bottom.

sinker: a heavy steel bar used in 'cable tool drilling' *(see drilling – cable tool)* and run above the bit and jars *(see jar)* to provide weight for operating the jars.

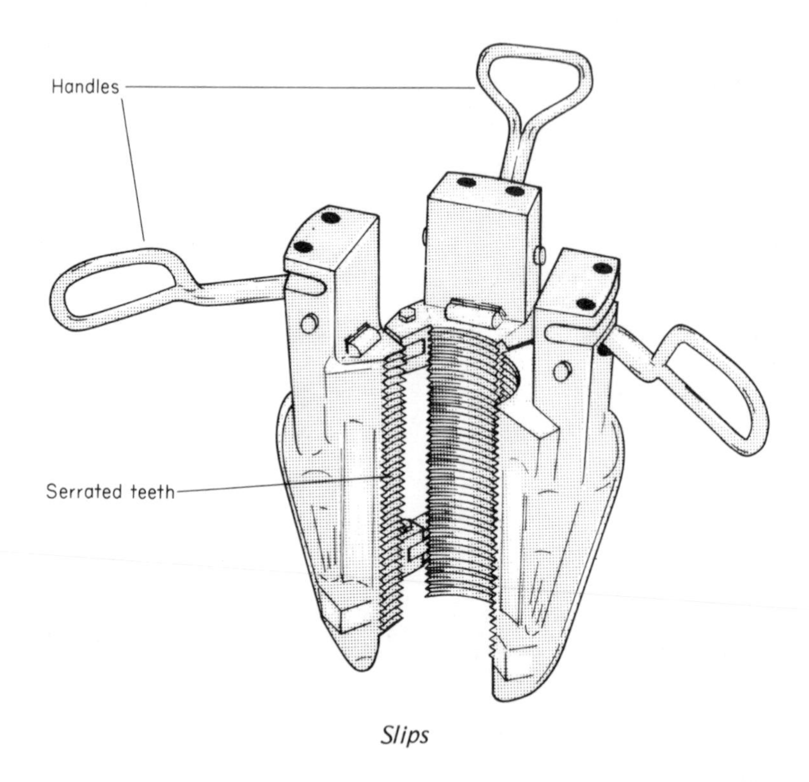

Slips

slips: steel wedges fitted with replaceable teeth like inserts which drop into position in the rotary table master bushings to hold the drill pipe or casing in the rotary table when connecting or disconnecting joints of pipe to the string.

slotted liner completion: when a producing sand is loose but consists of coarse material a length of casing having narrow vertical slots is placed through the reservoir formation and hung by means of a liner hanger in the final production casing string. The slots in the liner prevent particles of the reservoir rock from entering the production casing string.

slurry: a mixture of water and special grade cement which is pumped into the well to cement a casing string *(see casing string)* or to plug off a loss circulation zone *(see lose returns)* or to 'plug back' *(see plug back)* a bore hold.

Different grades of cement have properties which will accommodate down hole temperature conditions which increase according to the depth of a hole and they also provide a choice of initial setting times which is a most important consideration when cementing long strings of casing. It will be appreciated that the cement slurry being pumped to cement a casing string must not take its initial set before the entire volume of slurry required has been pumped.

slush pit: pit used for storing drilling mud *(see mud).*

slush pump: piston type pump with changeable liners and pistons used for handling the circulating fluid. By changing liners and pistons these pumps are able to accommodate varying pressure and volume discharge situations *(see Duplex pumps).*

snubber: device used for forcing drill pipe into a hole which is under pressure or to recover pipe under similar conditions. A snubber will usually incorporate hydraulic rams erected over the rotary table *(see rotary table).*

sour gas: gas which usually has an unpleasant odour and often contains hydrogen sulphide.

spacing: the distance between production wells drilled into the same reservoir.

spaghetti: tubing or drill pipe of a very small outside diameter (below, say, 2 inches – 51 mm).

spear: a 'fishing tool' *(see fishing tools)* used to enter the bore of a lost drill string.

spider: a heavy duty steel housing fitted with wedged type slips which is used to support casing being run into a hole.

spinning chain: chain which is wrapped around the drill pipe and operated from the drawworks *(see drawworks)* to speed up the operation of breaking out and 'making up' lengths of pipe when 'round tripping' *(see round trip).*

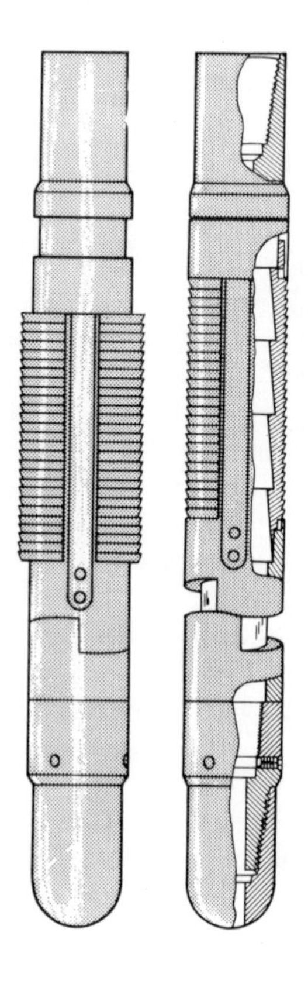

Spear "Fishing Tools"

spud in: the operation of drilling the first foot of a new hole.

squeeze job: pumping cement slurry into a porous formation under pressure.

stab: the operation of guiding one end of a pipe into the connection of another pipe to 'make up' a connection.

stabbing board: a temporary platform set in the derrick *(see derrick)* about 30 ft (9.1 m) above the derrick floor to allow the derrick man to handle joints of casing when running a string of casing *(see casing string)* into a hole.

stabilizer: a tool run above the bit *(see rock bit)* in a drilling string to assist in the drilling of a directional hole *(see directional drilling)* or to maintain the direction of a vertical hole.

stand: a length of drill pipe usually made up of three 'singles' *(see single)*

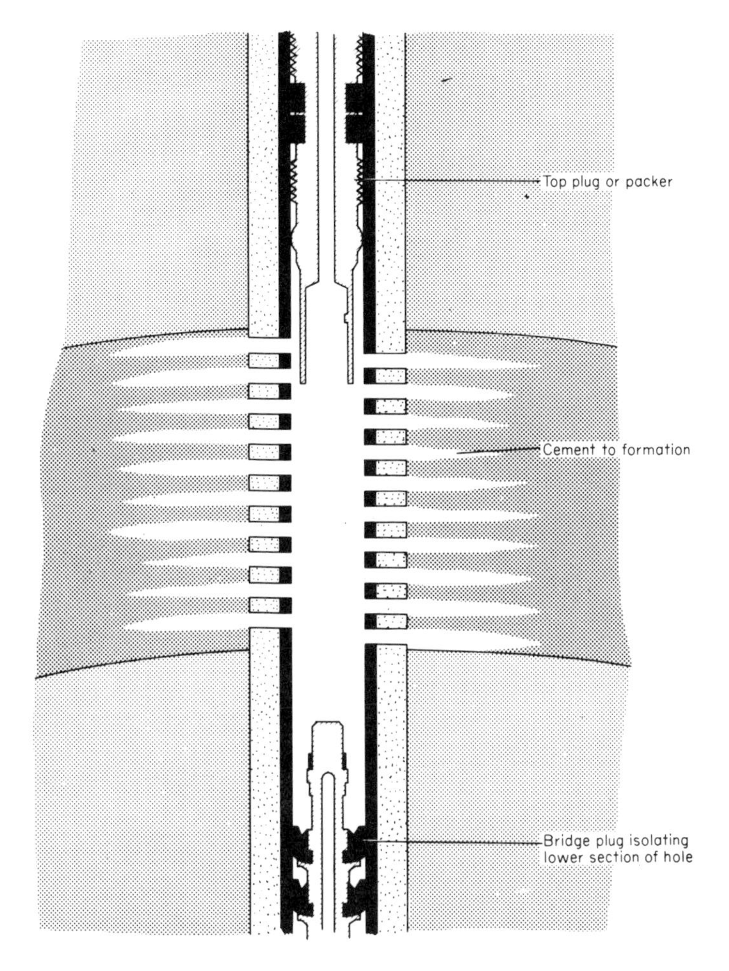

Squeeze Cementing

and about 90 ft (27.4 m) in length which is stacked in the derrick when pulling the pipe out of the hole.

stand pipe: a steel pipe located in the derrick *(see derrick)* alongside one leg and to which the flexible kelly hose *(see hose – rotary)* is attached to connect the slush pump *(see slush pump)* discharge line to the kelly hose and so to the 'swivel' *(see swivel).*

step out well: well drilled away from a discovery well to assess the reservoir area.

stove pipe: riveted type of casing which was used in old cable tool holes *(see drilling – cable tool).*

straddle test: the technique of setting down hole packers above and below a reservoir formation to conduct a 'flow test' *(see flow test).*

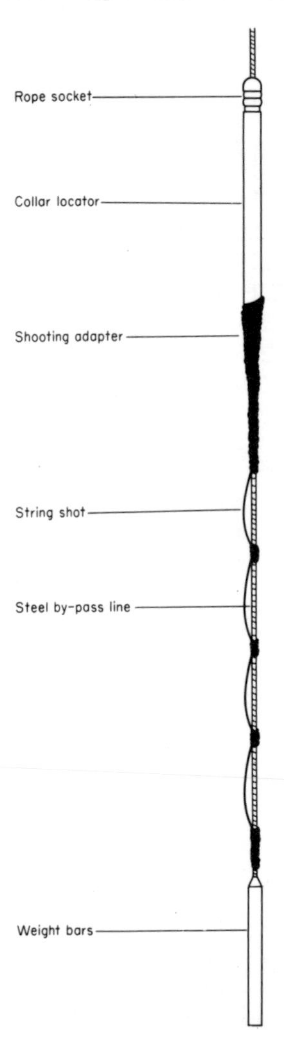

String shot assembly

string shot: a charge run inside the drill pipe on an electric cable which serves to blow off the pipe above an obstruction when the drill string is stuck in the hole and thus enables the free section of the string to be recovered and at the same time simplifies 'fishing' *(see fish)* the pipe remaining in the hole.

strip a well: to pull out rods and tubing from a well at the same time. The term derives from the necessity to 'strip' the tubing over the rods a joint at a time.

stripper: a well which will only produce a very small amount of oil.

stuck pipe: drill pipe, casing, or tubing, which becomes stuck in a hole and cannot be raised or lowered and frequently cannot be rotated or circulated.

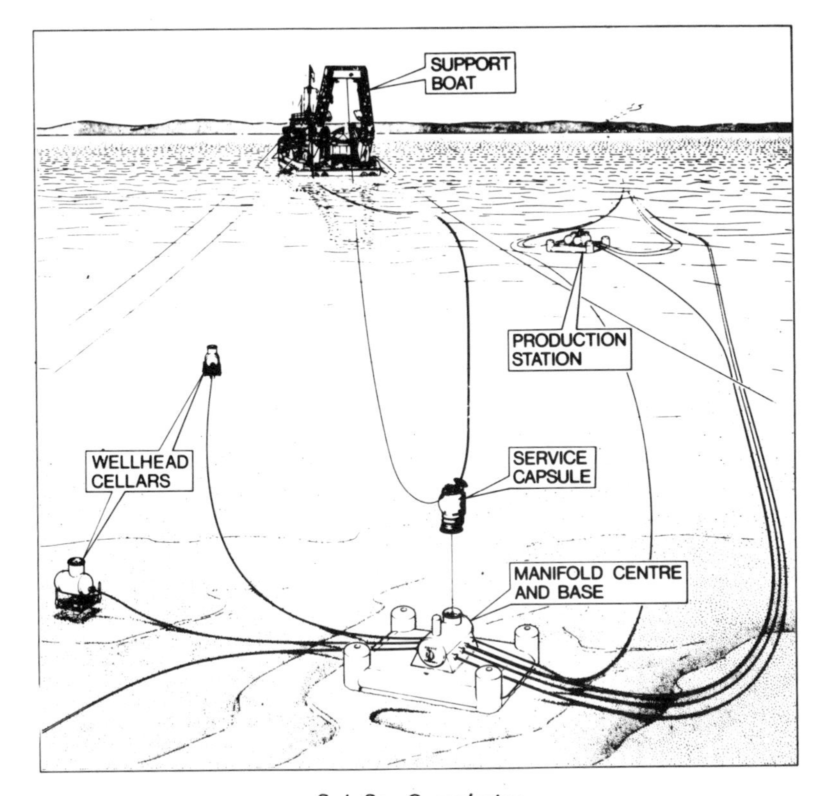

Sub-Sea Completion

sub-sea completion: the method of completing production wells whereby the well heads are located on the sea floor as opposed to the position of the Christmas trees *(see Christmas tree)* above sea level under the deck of a production platform. The oil or gas from a sub-sea completion is piped from the well heads to a fixing platform, a loading buoy, or to shore for processing.

Sub-sea installations are already in operation in shallower water conditions than exist in most parts of the North Sea.

substitute or 'sub': a screwed connection joint for matching pipes having different thread conditions.

substructure: the structure on which the derrick *(see derrick)* engines and drawworks *(see drawworks)* are installed.

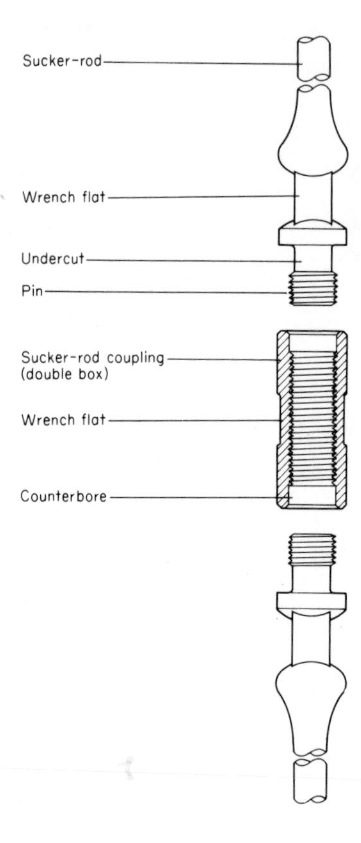

Sucker rod joints

sucker rods: rods, about 20 ft (6.1 m) long, which make up a string to operate a down hole pump *(see down hole pump).*

suction bailer: tubular device fitted with a foot valve which is used for bailing mud or fluid from a well usually in order to stimulate the production of reservoir fluids by reducing the hydrostatic head.

surface string: the first string of casing set in a well to shut off surface formations and act as an anchor for the drilling well head and blow out preventer stack *(see blow out preventer stack).*

swab: a tool fitted with rubber cups and run on a wire line which is used for removing fluid from a well bore and thus to reduce the hydrostatic head usually for the purpose of inducing the well to flow under formation pressure.

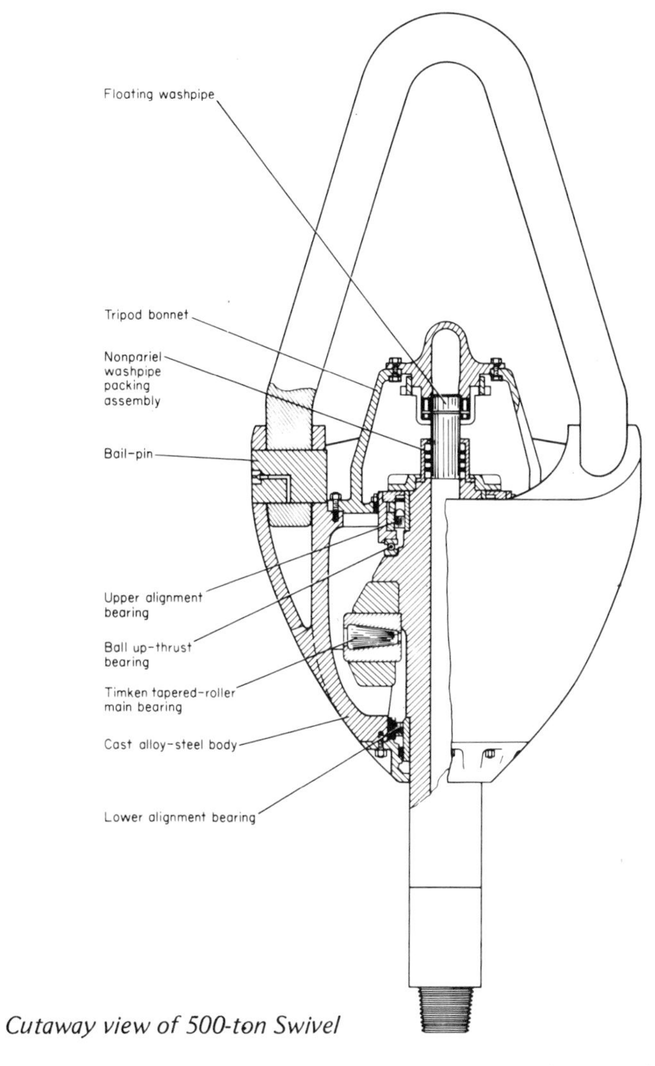

Cutaway view of 500-ton Swivel

swivel: as assembly which is suspended from the rig hook *(see hook)* and screwed to the top end of the kelly *(see kelly)* by a lefthand thread connection. The swivel supports the drilling string and allows rotation of the pipe and provides a means of circulating the drilling mud from the slush pump *(see slush pump)* discharge to the drill pipe via the kelly hose *(see hose – rotary).*

syncline: a saucer shaped configuration of the formation *(see geology).*

T

tail board: the rearside extension of a 'walking beam' *(see walking beam).*

tail out rods: pulling the lower end of a sucker rod *(see sucker rods)* away from the well when laying down rods.

tail pipe: length of pipe which is run below a packer *(see anchor).* The tail pipe sets on the bottom of the hole and permits the weight of the pipe above the packer to be used to close the packer at a pre-determined point by lowering weight on to the packer element.

tank farm: a storage area consisting of an assembly of very large storage tanks with related pipe manifolds and pumping units.

tank strapper: the operator who measures the level of liquid in a tank to establish the contents.

tap (fishing): a tapered thread cutting tool which is run to enter and connect to a pipe which is lost in a hole.

tearing down: dismantling a rig and equipment on completion of a well.

telescoping derrick: a portable mast whose sections 'nest' inside each other and can be extended and erected by means of wire line tackle or hydraulics.

temperature survey: the running of an instrument which records the change of temperature in a hole and is used to locate the correct cementation of a casing string *(see casing string)* or to locate the inflow of water into a bore hole.

thief formation: *(see lose returns).*

thrible: a drill pipe stand *(see stand)* made up of three 'singles' *(see single)* and about 90 ft (27.4 m) in length.

throwing the chain: the technique of 'flipping' the spinning chain *(see spinning chain)* from the joint of pipe sitting in the rotary table to the joint of pipe being made up to the string. The chain is coiled with several wraps around the pipe and the power for spinning the pipe is upplied by the draw-works cathead *(see cathead).*

tight formation: a reservoir formation which has poor porosity or permeability conditions and does not allow free flow of fluid from the formation into the hole.

tong line: wire or rope used for pulling on the pipe tongs used for making up and breaking out sections of the drill string above the rotary table *(see rotary table).*

tongman: floorman who handles the tongs which are used for making up or breaking out joints of pipe from the drilling string.

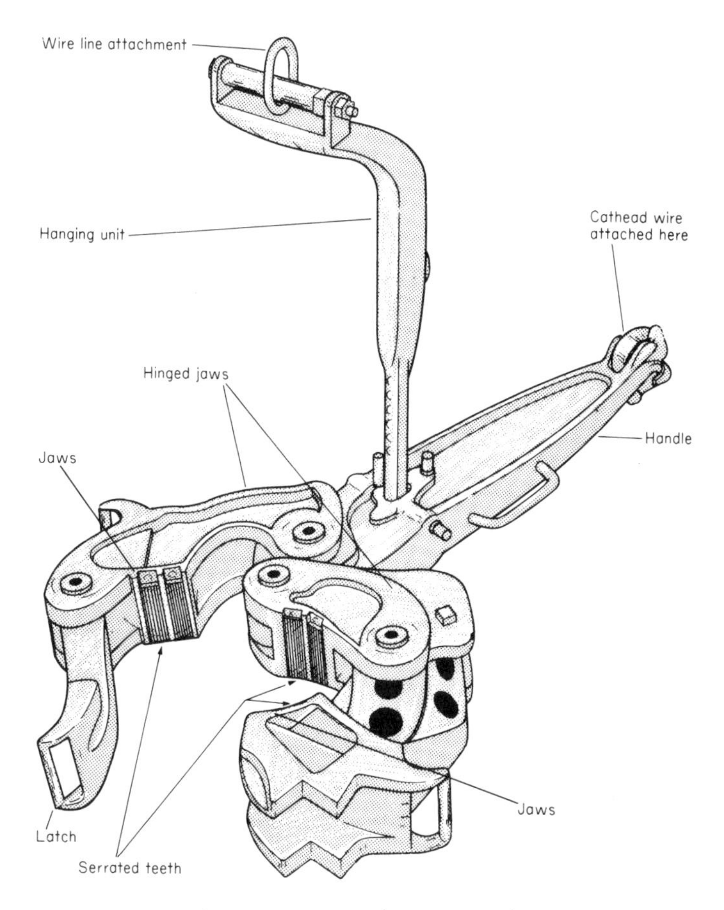

Large pipe tongs (or wrenches)

tongs: large pipe wrenches which hang in the derrick from wire lines and are used for making up or breaking out lengths of drill pipe, casing or tubing.

tongs (power): pneumatically or hydraulically operated tools which spin the drill pipe and take the place of the normal hand operated tongs *(see tongs).*

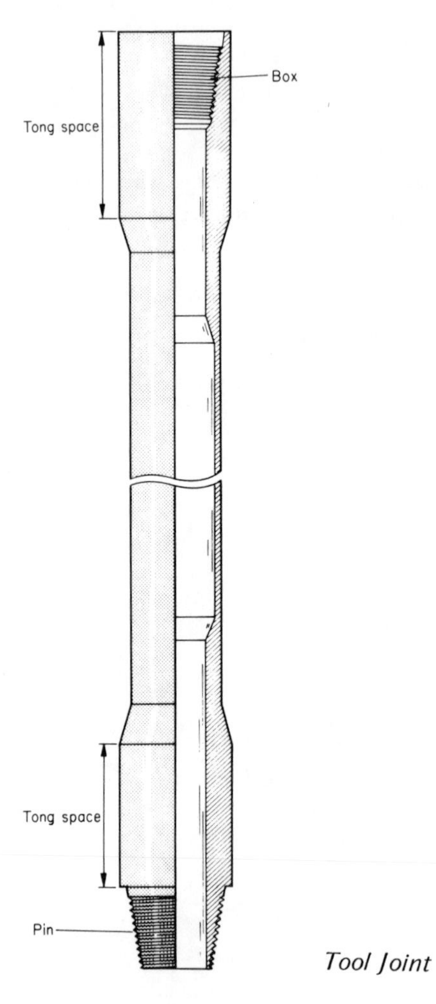

Tool Joint

tool joint: very heavy special steel couplings which are screwed or flash welded to the end of drill pipe 'singles' *(see single)* to facilitate joining the individual lengths to make up a drill string. Tool joints have coarse tapered threads which permit frequent coupling up and breaking out of joints and matching faces of 'pin and box' *(see drill pipe)* tool joints to prevent the escape of the mud circulating fluid.

tool pusher: (see *drilling crew).*

tour: shifts worked by the drilling crews and often known as a 'tour'. The tour may be of eight hours duration or of twelve hours duration and is usually referred to as: morning tour, evening tour and graveyard tour (0800 – 1600 hrs, 1600 hrs – midnight, midnight – 0800 hrs).

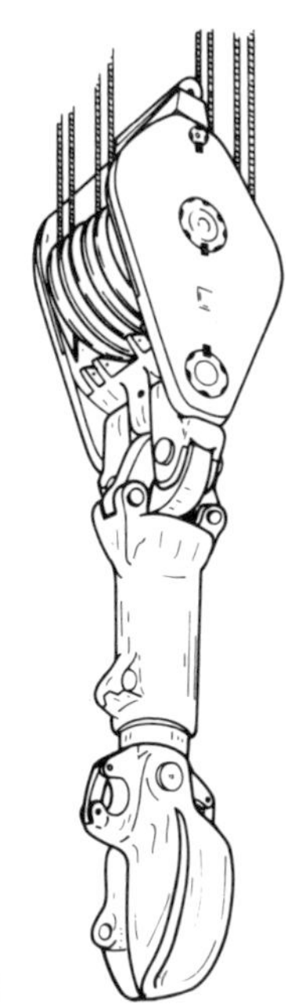

Travelling Block and Hook

travelling block: the block containing sheaves which is attached to the hook and accommodates the hoisting line via the derrick crown block to provide the means of hoisting or lowering the drill pipe and casing loads involved in a drilling operation.

tricone bit: a rock bit having three cutting cones mounted on high duty roller bearings.

trip: *(see round trip).*

trip gas: gas which enters the hole when a round trip *(see round trip)* is being made and is often 'pulled in' from the formation by the piston effect of pulling the pipe and bit at high speeds.

tubing: small size pipe of outside diameter $2\frac{3}{8}$ in to $2\frac{7}{8}$ in (60 mm to 73 mm) which is run into a production well to handle the fluid flow from the formation to surface.

tubing job: the operation of pulling or running tubing *(see tubing).*

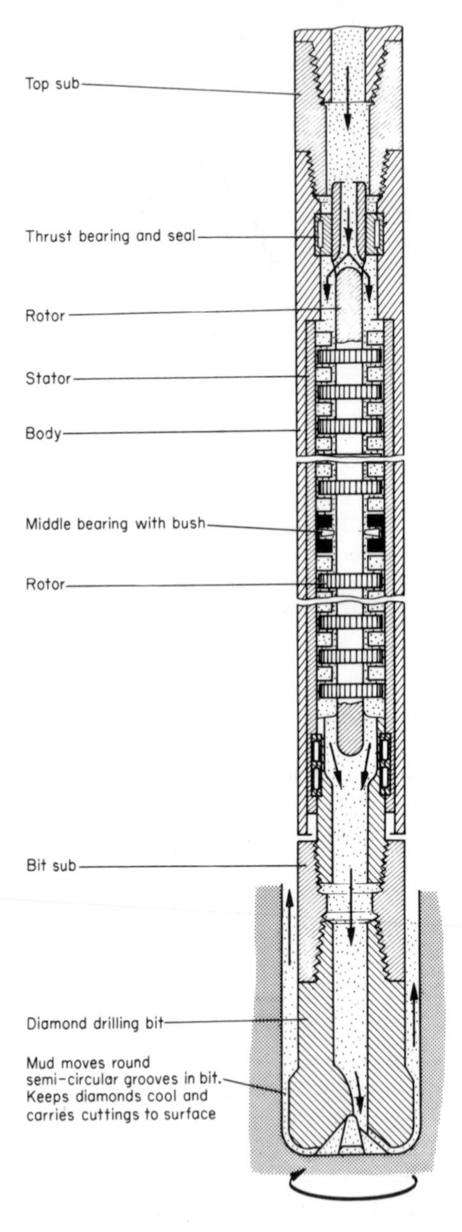

Turbo Drill

turbo drill: turbine situated at the bottom of the drill string which is operated by the mud circulation and turns the rotary bit without the drilling string having to be rotated by the rotary table *(see rotary table)* in the conventional manner. This tool is frequently used in directional drilling and using a diamond bit *(see rock bit).*

twist off: a failure of the drill string due to excess tortional stress or faulty pipe.

U

under ream: the operation of enlarging a hole below a casing shoe *(see casing shoe)* by means of using an 'under reamer' *(see ream)* tool fitted with expanding cutter blades which are actuated by the circulating fluid pressure.

V

V. door: opening in the derrick structure at floor level to permit the pulling in of lengths of drill pipe or casing from the racks outside the derrick.

vibrating screen: unit which is used for removing cuttings from the circulating mud on its return from the well annular space. The returning mud flows through a fine wire mesh which is vibrated by an electric motor drive and causes the formation cuttings to be separated from the circulating fluid and discharged into a collecting pit whilst the mud returns through a ditch or fluming to the slush pumps for re-circulation in the well.

viscosity: the measure of a liquid's resistance to flow and is an important factor in the properties of a circulating mud *(see marsh funnel).* A heavy mud is not necessarily a viscous mud.

W

waiting on cement (w.o.c.): the expression used for the period during which a rig is shut down to allow cement to set after a casing string *(see casing string)* has been cemented in the hole or a cement plug *(see cement plugs)* has been placed.

walking beam: an oscillating beam supported by the sampson post *(see sampson post)* from which on a cable tool rig *(see drilling – cable tool)* the drilling tools are suspended on a wire line in the hole and operated. In a pumping well the 'walking beam' actuates the down hole pump *(see down hole pump)* by means of a string of rods *(see sucker rods).*

wall hook: a 'fishing tool' *(see fishing tools)* used to pull the top of a 'fish' *(see fish)* into the centre of the hole to enable an overshot *(see overshot)* or similar device to be connected to recover the 'fish'.

wall scraper: tool used for scraping the wall of a hole to remove 'mud cake' *(see mud – cake)* opposite a 'thief formation' *(see lose returns)* or some similar intruding deposit.

washout: enlargement in a well bore due to erosion by the circulating fluid or due to the solubility of a salt formation causing large 'caves'.

washover: the operation of working a pipe fitted with a cutter head over a 'stuck pipe' by rotating and circulating in order to remove the obstruction causing the 'fish' *(see fish)* to be stuck.

water (bottom): water entering the hole from below the oil in a producing sand or from a sand below the producing sand.

water (connate): water which was trapped in the formation when it was deposited many millions of years ago.

water drive: the situation whereby a well is caused to flow oil by water pressure on the reservoir due to a head of water existing in the formation.

water flooding: the process of injecting water into a reservoir formation for the purpose of flooding out the oil towards a production well. Special injection wells are drilled for this purpose.

water head: pressure caused by a column of water – approximately .434 lbs/in^2 (.033 kg/cm^2) per foot of water head.

water string: a string of casing *(see casing string)* the purpose of which is to shut out formation water.

water table: the platform at the top of a derrick on which the 'crown block' *(see crown block)* is located.

weight indicator: an instrument mounted in sight of the driller which records the weight of the drilling string or casing string *(see casing string)* hanging from the travelling block *(see travelling block)* and hook *(see hook)* assembly. This unit provides the driller with information as to the weight he is applying to the bit during the drilling operation or the pull which he is

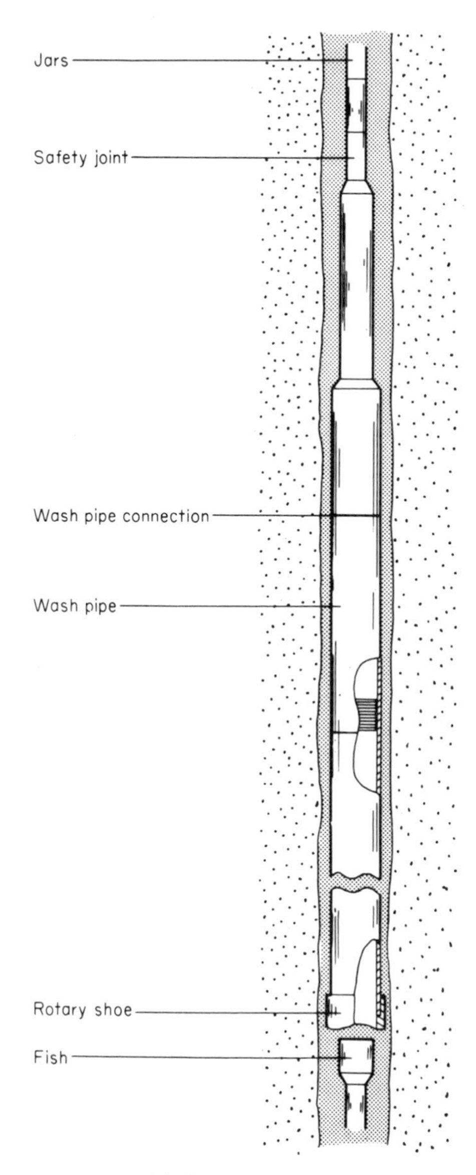

Washover operation

applying to a drill string or casing string when hoisting or pulling or lowering through 'tight hole' conditions. The load factor is transmitted to the weight indicator by the diaphragm attached to the dead line anchor *(see anchor – dead line).*

weighting material: heavy material such as barium sulphate which is added to the circulating fluid to increase its specific gravity for the purpose of controlling high pressures encountered when drilling into reservoir formations or to support formations which tend to 'cave'.

well: the word 'well' may conjure up thoughts of shallow hand dug holes put down for the purpose of obtaining water. A modern oil or gas well bears very little relationship to this conception. Apart from holes which are drilled to obtain geological information the usual purpose for drilling a well is to provide a conduit for the recovery of oil, gas, or water from an underground reservoir formation to the surface.

There is a saying in the drilling industry that 'no two wells are the same' and this is very true in that holes may vary in depth from a few hundred feet to 30 000 feet or more. Geological conditions and types of formation encountered in holes, even small distances apart may, and do, vary tremendously. The final resulting target may prove to be oil, water, gas, steam or even nothing at all and the pressures of the fluid in the reservoir may vary between a few pounds per square inch to many thousands of pounds per square inch. Some holes will be well consolidated and give little trouble to drill whereas others may encounter difficulties which require the use of very special and expensive muds and in some cases may present insurmountable problems.

Before the well is 'spudded in' *(see spud in)* the geologist and geophysicist can give a calculated estimate of the conditions which should be encountered during drilling but, unfortunately, such forecasts are rarely, if ever, entirely accurate and, consequently, the drilling engineer is continually presented with problems which cannot be foreseen until they are actually encountered. Planning of the entire drilling operation for each well must depend upon information available from geological and geophysical surveys, from results when drilling previous wells in the same or similar areas, and with a knowledge of equipment and materials which are in existence to handle the possible problems The entire drilling operation for any particular well depends upon the casing programme *(see casing string)* which is designed before drilling commences.

The capacity of the rig *(see rig – drilling)* and the horse power of the engines required is dictated by the casing loads which will have to be handled. The drill pipe and drill collar programme and the diameter of bits required to carry out the programme must be set up before 'spudding' the well in.

Forward planning will have to provide for the immediate availability of the cement required for cementing all the casing strings to be run and allowance must be made for cement plugs *(see cement plugs)* which may have to be run during the drilling operation. Mud materials must be readily available *(see mud)* to mix the most suitable fluid to handle geological and pressure conditions encountered as the well progresses and many of these conditions cannot be foreseen until they actually occur.

Fishing tools *(see fishing tools)* to accommodate any possible failure of down hole equipment or unforeseen circumstances must be readily available. Well head fittings and safety devices such as blow out preventers *(see blow out preventer stack)* will be designed to accommodate any possible pressure conditions which could occur and a large safety factor has to be provided to allow for unforeseen conditions especially when drilling a 'wild cat' *(see wild cat).*

In remote areas of the world, provision of all the services and supplies required to drill a well, to provide accommodation and to feed the personnel and to allow for unexpected situations presents a very formidable and expensive problem for management. Many a well has cost £1,000,000 or more to drill and many such wells have proved to be 'dusters' *(see duster).* Of course, it will be appreciated that the difficulties to be overcome and the magnitude of the problems to be tackled increases very considerably when the well is in deep water and rough conditions such as are experienced in the North Sea.

well log: *(see log – well).*

wet gas: *(see gas – wet).*

whipstock: a steel wedge about 20 ft (6.1 m) long which is set in a hole and serves to 'kick off' the hole to start a directional drilling operation.

widow maker: anything which could cause the death or serious injury of a workman.

wild cat: a well drilled in an area which is unproven. On average one 'wild cat' in ten or twelve wells may prove productive. The drilling of a 'wild cat' demands special safety precautions as the down hole formation and pressure conditions are unknown and may present quite unexpected and dangerous problems.

wildcatter: an operator who drills in the hope of discovering oil or gas in an area which is not proved to be an 'oil field' (i.e. a productive area already established.

wild well: a well which is out of control and blowing fluid or gas from the down hole reservoir. This condition, which is fortunately a rare occurrence, is serious if not disastrous. In the event of a 'blow out' *(see blow out)* of this nature specialists are called in to tackle the problem and no expense is spared to bring the 'wild well' under control. Every case is different but the aim is to mud off the well, i.e. contain the flow by pumping mud into the hole or to seal the hole by the use of an explosive or, in extreme cases, to drill a directional hole from some remote point into the 'wild hole' in order to inject mud or cement.

window: an opening milled in a casing joint to permit drilling a new hole by directional methods away from the original hole.

work over job: carrying out a remedial operation on a production well such as 'plugging back' *(see plug back)*, pulling a liner for cleaning, squeeze cementing *(see cement squeeze)* or fracturing *(see fracturing – formation)* etc.

X

Xmas tree: *(see Christmas tree).*

APPENDICES

APPENDIX 1

World Refining and Production Output – 1938–74

Table I: REFINING[1]

Thousand tonnes

	1938	1950	1965	1974[2]
W. EUROPE	**15,950**	**46,220**	**408,885**	**939,775**
Inc. Italy[3]	2,100	5,900	100,275	200,125
W. Germany	2,400[4]	4,700	80,180	144,210
United Kingdom	1,900	11,450	72,190	148,520
France	7,600	15,700	69,275	167,150
Netherlands	800	5,600	31,705	91,225
MIDDLE EAST	**14,000**	**48,750**	**99,970**	**143,315**
Inc. Iran	11,400	25,100	25,355	31,100
Kuwait	–	1,250	17,125	27,600
Saudi Arabia	–	7,000	12,500	21,300
Bahrain	1,600	7,750	10,550	12,500
AFRICA	**950**	**950**	**26,370**	**64,340**
NORTH AMERICA	**234,100**	**369,500**	**567,590**	**837,040**
Inc. USA	225,400	350,400	512,395	742,270
Canada	8,700	19,100	55,195	94,770
LATIN AMERICA	**39,800**	**73,810**	**199,295**	**362,395**
Inc. Venezuela	2,700	14,200	54,130	75,640
Netherlands Antilles	21,400	33,100	36,000	42,500
Argentina	4,900	7,600	19,525	31,590
Mexico	4,900	8,900	23,000	35,600
FAR EAST AND AUSTRALASIA	**13,700**	**14,680**	**165,740**	**453,000**
Inc. Japan	2,600	2,600	94,790	245,710
Australia	200	900	20,850	32,950
India	1,900[5]	290	11,645	26,220
Indonesia	8,000	8,650	13,860	21,310
SINO-SOVIET AREA	**45,200[6]**	**49,190**	**267,150**	**563,000**
Inc. USSR	31,500	37,500	225,000	422,000
WORLD TOTAL	**363,700**	**603,100**	**1,735,000**	**3,362,865**

NOTES:

1 *As at year end*
2 *Preliminary*
3 *Including reserve capacity*
4 *All Germany*
5 *Includes Pakistan and Burma*
6 *Includes countries which entered the USSR Bloc later*

APPENDIX 1 (continued)

Table II: PRODUCTION[1]

Thousand tonnes

	1938	1950	1965	1974[2]
W. EUROPE	**690**	**1,980**	21,140	20,350
Inc. W. Germany	550[3]	1,120	7,800	6,200
MIDDLE EAST	**16,200**	**88,420**	**422,320**	**1,083,240**
Inc. Iran	10,400	32,260	93,740	301,040
Saudi Arabia	100	26,620	100,950	409,040
Kuwait	–	17,290	108,730	114,370
Iraq	4,400	6,650	64,360	95,550
Abu Dhabi	–	–	13,560	
Neutral Zone	–	–	18,950	28,590
AFRICA	–	**180**	**100,650**	**274,150**
Inc. Libya	–	–	58,700	74,030
Algeria	–	80	26,480	50,140
Nigeria	–	–	13,380	111,900
NORTH AMERICA	**171,600**	**289,080**	**477,700**	**592,690**
Inc. USA	170,700	285,200	433,480	498,620
Canada	900	3,880	44,220	94,070
LATIN AMERICA	**44,230**	**102,550**	**241,735**	**250,280**
Inc. Venezuela	28,100	78,140	181,120	154,900
Mexico	5,500	10,490	18,600	30,950
Argentina	2,400	3,460	14,050	21,580
Trinidad	2,600	2,980	6,950	9,410
Columbia	3,100	4,850	10,560	9,050
FAR EAST AND AUSTRALASIA	10,000	11,830	33,100	110,610
Inc Indonesia	7,400	6,450	24,100	69,130
SINO-SOVIET AREA	**37,780[4]**	**44,430**	**268,290**	**537,580**
Inc. USSR	30,100	37,500	243,000	456,940
Rumania	6,900	4,100	12,550	14,750
WORLD TOTAL	**280,500**	**538,470**	**1,564,935**	**2,868,900**

NOTES;

1 *Crude Oil and natural gas liquids*
2 *Preliminary*
3 *All Germany*
4 *Includes countries which entered USSR Bloc later*

APPENDIX 2

Miscellaneous world statistics

Table I: PROVEN RESERVES OF CRUDE OIL AND NATURAL GAS LIQUIDS

Thousand tonnes

	1973	1974*	% Share of total
WORLD TOTAL	**86,583,330**	**98,539,060†**	**100.00%**
of which			
MIDDLE EAST	**47,967,290**	**55,322,930**	**56.1%**
Abu Dhabi	2,945,200	4,109,580	
Bahrain	49,310	46,030	
Dubai	342,470	331,510	
Iran	8,219,160	9,041,080	
Iraq	4,315,060	4,794,510	
Israel	270	270	
Kuwait	8,767,100	9,972,580	
Neutral Zone	2,397,260	2,369,860	
Oman	719,180	821,920	
Qatar	890,410	821,910	
Saudi Arabia	18,082,150	22,534,200	
Sharjah	205,480	205,480	
Syria	972,600	205,500	
Turkey	61,640	68,500	
SINO-SOVIET AREA	**14,109,600**	**15,260,300**	**15.6%**
AFRICA	**9,239,670**	**9,356,050**	**9.5%**
Algeria	1,046,570	1,054,790	
Angola/Cabinda	205,480	160,960	
Congo Republic	669,590	667,670	
Egypt	702,050	506,850	
Gabon	205,480	239,720	
Libya	3,493,140	3,643,830	
Morocco	100	60	
Nigeria	2,759,720	2,863,000	
Tunisia	130,140	150,680	
Zaire	27,400	68,490	

* *Preliminary*

† *= 13.8% increase over previous year*

Continued

Appendix 2 – Table I (continued)

Proven reserves *Thousand tonnes*

	1973	1974*	% Share of total
WORLD TOTAL	**86,583,330**	**98,539,060†**	**100.00%**
of which			
NORTH AMERICA	**6,600,000**	**6,625,000**	**6.7%**
USA	5,400,000	5,425,000	
Canada	1,200,000	1,200,000	
LATIN AMERICA	**4,334,390**	**5,555,360**	**5.6%**
Argentina	342,470	321,370	
Barbados	150	150	
Bolivia	35,610	34,250	
Brazil	109,450	106,160	
Chile	16,990	27,400	
Colombia	196,160	123,290	
Ecuador	777,400	342,470	
Mexico	493,150	1,860,540	
Peru	143,840	342,470	
Trinidad	301,370	342,470	
Venezuela	1,917,800	2,054,790	
WESTERN EUROPE	**2,190,600**	**3,536,150**	**3.6%**
Austria	21,500	24,930	
Denmark	34,040	33,830	
France	10,960	19,450	
West Germany	74,520	75,340	
Italy	29,040	102,740	
Netherlands	34,380	34,250	
Norway	547,940	1,000,000	
Spain	8,220	40,140	
United Kingdom	1,370,000	2,150,680	
Yugoslavia	60,000	54,790	

* *Preliminary*

† *= 13.8% increase over previous year*

Continued

Appendix 2 – Table I (continued)

Proven reserves *Thousand tonnes*

	1973	1974*	% Share of total
WORLD TOTAL	86,583,330	98,539,060†	100.00%
of which			
FAR EAST AND AUSTRALASIA	2,141,780	2,883,270	2.9%
Afghanistan	12,320	11,640	
Australia	315,070	315,070	
Brunei/Malaysia	219,180	342,470	
Burma	9,320	8,900	
India	106,710	114,250	
Indonesia	1,438,360	2,054,790	
Japan	2,600	4,110	
New Zealand	30,680	10,270	
Pakistan	4,250	3,970	
Taiwan	1,920	2,050	
Thailand	1,370	15,750	

* *Preliminary*

† *= 13.8% increase over previous year*

Source: Oil & Gas Journal, Year End number, except USA and Canada (estimated)

Table II: WORLD TANKER FLEET – at end of 1974

	No.	D.W. tons	% Share of total
WORLD TOTAL	**3,638**	**255,769,852**	**100.00%**
of which			
Liberia	886	73,961,708	28.92%
United Kingdom	398	32,155,797	12.57%
Japan	220	29,184,002	11.41%
Norway	241	23,892,409	9.34%
USA	303	10,368,746	4.05%
Others	1,590	86,207,190	33.71%

Table III: ENERGY CONSUMPTION

	Oil	Coal*	Natural Gas	Nuclear Power & Hydro-electricity
FREE WORLD				
1966	47.5%	27.7%	16.3%	8.5%
1969	49.5%	24.2%	18.0%	8.3%
1974	51.0%	19.0%	20.0%	10.0%
UNITED KINGDOM				
1966	37.5%	58.7%	0.4%	3.4%
1969	42.7%	50.7%	2.7%	3.9%
1974	41.5%	35.0%	15.8%	4.1%

**Based on 1 ton of coal = 0.625 ton oil equivalent*

Table IV: TYPICAL PERCENTAGE YIELD PATTERNS

	Venezuela Medium	Kuwait	Arabian Light
Balance	4%	5%	5%
Gasolines	15%	19%	21%
Middle distillates	19%	26%	30%
Residual fuel oil	62%	50%	44%

CONVERSION FACTORS

1 Long Ton	=	1.01605 Tonnes
1 Tonne	=	0.98421 Long Ton
1 Barrel	=	34.97261 Imperial Gallons
100 million cubic feet of natural gas per day	=	approximately 360 million therms per annum
	=	700,000 tons per annum liquid

Approximate Barrels/Tonnes

Crude Oils	8.0–6.6	Aviation Spirit	9.1–8.2
Motor Spirit	9.0–8.1	Kerosines	8.2–7.6
Gas Oils	7.8–7.1	Diesel Oils	7.8–6.9
Lubricating Oils	7.5–6.7	Fuel Oils	6.9–6.5

(Reproduced by courtesy of The Institute of Petroleum)

APPENDIX 3

Britain's Petroleum Industry

Table I: UK REFINING CAPACITY

Thousand tonnes

	1938	1950	1965	1974
TOTAL	1,900	11,450	72,190	148,520
of which:				
ESSO				
Fawley	700	1,100	11,500	19,500
Milford Haven	–	–	6,300	15,000
SHELL				
Stanlow	–	1,200	10,350	18,000
Shell Haven	–	2,000	9,350	10,000
Teesport	–	–	–	6,000
Heysham	–	1,800	1,950	2,200
Ardrossan	–	150	175	200
BRITISH PETROLEUM				
Kent	–	–	9,500	10,900
Llandarcy	360	2,850	7,800	8,300
Grangemouth	360	1,750	4,500	8,700
Belfast	–	–	1,300	1,500
Pumpherston[1]	150	160	–	–
TEXACO				
Pembroke	–	–	5,100	9,000
PHILLIPS-IMPERIAL				
Billingham, Cleveland	–	–	1,000	5,000
LINDSEY				
Killingholme	–	–	–	9,250
MOBIL				
Coryton	–	–	2,400	9,000
GULF				
Milford Haven	–	–	–	5,000
CONTINENTAL OIL				
South Killingholme	–	–	–	4,500
AMOCO (U.K.)				
Milford Haven	–	–	–	4,000
PHILMAC OILS				
Eastham	–	–	–	600

1 *Closed down 1964*

Continued

Appendix 3 – Table 1 (continued)

UK Refining capacity

Thousand tonnes

	1938	1950	1965	1974
TOTAL of which:	1,900	11,450	72,190	148,520
BURMAH OIL TRADING				
Ellesmere Port	100	120	250	1,500
Barton Trafford Park[2]	100	130	175	–
BERRY WIGGINS				
Kingsnorth	70	95	285	285
Weaste[2]	60	70	170	–
WILLIAM BRIGGS				
Dundee	–	25	85	85

2 *Closed down 1972*

APPENDIX 3 (continued)

Table II: UK CONSUMPTION*

Thousand long tons

	1938	1950	1965	1974
Aviation spirit	113	458	167	53
Aviation turbine fuels	–		2,220	3,693
Motor spirit (inc. motor benzole)	4,830	5,195	10,739	16,223
Industrial spirits (inc. industrial benzole)	38	81	234	70
White spirit	72	149	148	134
Kerosine	721	1,363	1,705	2,764
of which:				
Burning oil	543	562	1,585	2,738
Vaporising oil	178	801	120	26
Derv fuel	387	1,034	3,844	5,432
Gas, diesel & fuel oils	1,608	5,338	38,319	56,937
of which				
Gas/diesel oil	797	1,595	6,856	13,366
Fuel oil	628	3,093	27,296	36,735
Refinery Consumption	183	650	4,167	6,856
Lubricating oils and greases	564	749	1,106	1,029
Paraffin wax and scale	48	43	56	88
Propane and butane	2	30	1,004	1,366
Bitumen	607	621	1,460	2,205
Chemical feedstock (other than naphtha)	–	–	475	500
Naphtha/LDF	–	216	4,919	7,579
Other products	–		524	153
TOTAL	**8,990**	**15,277**	**66,920**	**98,226**

**Excluding Bunkers for ships in foreign trade.*

APPENDIX 3 (continued)

Table III: UK IMPORTS

	1938	1950	1965	1974*
				Thousand long tons
Crude oil	2,272	9,246	65,653	111,758
Refined products	9,390	9,896	19,713	14,887
TOTAL	11,662	19,142	85,366	126,645

Sources as percent of total

	1938	1950	·1965	1974*
Middle East	24.2	54.1	46.1	64.9
Latin America	48.6	32.7	17.2	3.3
Africa	–	–	22.3	14.4
North America	18.0	4.7	0.5	0.3
Other (mainly Europe)	9.2	8.5	13.9	17.1

Imports of crude oil by sources

Thousand long tons

	1938	1950	1965	1974*
Kuwait	–	3,580	14,175	16,979
Iraq	557	486	10,354	3,153
Libya	–	–	11,161	8,603
Venezuela	539	308	7,502	3,502
Iran	211	2,325	4,140	14,293
Nigeria	–	–	6,823	7,775
Trucial States	–	–[1]	3,467	10,226
Saudi Arabia	–	1,067	4,380	35,062
Netherlands	–	19	1,062	760
Netherlands Antilles	628	1,119	508	–
USA	58	74	49	–
Irish Republic	–	–	304	4,145
Others	279	268	1,728	7,260
TOTAL	2,272	9,246	65,653	111,758

*Preliminary

1 *Included in Kuwait for 1950*

APPENDIX 3 (continued)

Table IV: UK EXPORTS AND RE-EXPORTS

Thousand long tons

	1938	1950	1965	1974*
Crude oil and products	592	1,193	10,792	15,342
Bunkers	1,362	2,867	5,176	4,684

Exports and re-exports of crude oil and refined liquid products by destination

Thousand long tons

	1938	1950	1965	1974*
Sweden	No	93	3,262	2,854
Denmark	analysis	19	2,065	2,675
Western Germany	available	24	1,001	905
Norway	–	40	1,011	1,009
Netherlands	–	32	1,304	2,473
Irish Republic	–	328	491	2,983
France	–	13	347	35
Others	–	644	1,311	2,410
TOTAL	592	1,193	10,792	15,342

**Preliminary*

Table V: UK PRODUCTION – 1974

Crude oil (inc. North Sea condensate)	245,857 long tons
Natural gas	27,306,008 long tons

Table VI: NUMBER OF WELLS DRILLED OFFSHORE UK

	Exploration and appraisal	Development
1964	1	–
1969	52	27
1974	100	21

Table VII: COMPARATIVE RETAIL PRICES OF MOTOR SPIRIT

Date	Pump price	Tax	Tax as % of pump price
June 1950	3s.0d.	1s.6d.	50.0%
June 1955	4s.6d.	2s.6d.	55.6%
June 1960	4s.8d.	2s.6d.	53.6%
June 1965	5s.2d.	3s.3d.	62.9%
June 1970	6s.2d.	4s.6d.	73.0%
June 1975	72.5p	22.5p + 14.5p VAT	51.0%

(All statistics in this appendix reproduced by courtesy of the Institute of Petroleum)

APPENDIX 4 Northern North Sea Oil and Gas Discoveries (as at April 1975) UK SECTOR

Field	Block	Operator	Recoverable reserves mm bbls (10^6)	Estimated max. output thousand b/d	Discovery date	Development date and plans
Brent	211/29 3/4	Shell/Esso Texaco	2000	500	June 1971 October 1973	1977 SPAR + tanker loading initially
Dunlin	211/23 211/24	Shell/Esso Conoco etc	500	150	May 1973 November 1973	1977 These 5 fields to be
Cormorant	211/26	Shell/Esso	350	120	August 1972	1977 produced together via
Hutton*	211/28 211/27	Conoco etc Amoco	200	100	September 1973	1978 pipeline to Shetland; Cormorant Platform
Thistle	211/18 211/19	Burmah Conoco	400	150	September 1972	1977 as collecting platform.
Ninian	3/3 3/8	Chevron BP	2000	450	January 1974	1978 These 3 fields will
Alwyn*	3/14	Total	400	150	September 1972	1978 probably share a
Heather	2/5	Union	150	50	November 1973	1978 pipeline to Shetland.
Magnus*	211/12	BP	150	50	July 1974	1980 Tanker loading
Statfjord	211/24 33/9, 12	Conoco etc Mobil/Statoil	3000	200	May 1974	1978 SBM + tanker loading
Tern*	210/25	Shell/Esso			April 1975	Under appraisal
Piper	15/17	Occidental	650	220	January 1973	1976 Pipeline to Flotta
Claymore*	14/19	Occidental	350	110	May 1974	1977 (Orkney)
Beryl	9/13	Mobil	400	130	September 1972	1976-7 SBM + tanker loading

(continued)

APPENDIX 4 (continued)

Field	Block	Operator	Recoverable reserves mm bbls (10^6)	Estimated max. output thousand b/d	Discovery date	Development date and plans
Unnamed*	9/13 9/12	Mobil Union			August 1974 March 1975	Under appraisal
Unnamed*	9/28	Hamilton			August 1973	Under appraisal
Unnamed*	14/20	Texaco			February 1975	Under appraisal
Unnamed*	15/16	Texaco	1800	400	December 1974	Under appraisal
Forties	21/10	BP	1800	400	October 1970	1975 Pipeline to Cruden Bay
Andrew*	16/28	BP	200	100	October 1974	1980 Will be tied in
Maureen*	16/29	Phillips	400	150	February 1973	1978 with Forties
Montrose	22/17	Amoco	150	50	November 1971	1976 SBM + tanker loading
Auk	30/16	Shell/Esso	75	40	January 1971	1975 SBM + tanker loading
Argyll	30/24	Hamilton	50	15	August 1971	1975 SBM + tanker loading
Josephine*	30/13	Phillips			September 1970	Under appraisal
GAS						
Frigg	10/1 25/1	Elf/Total Petronord	212,000 million m^3	1.5 billion cfd	April 1972	1977 Pipeline to St Fergus
Lomond*	23/21	Amoco			April 1972	Under appraisal
Unnamed*	211/13	Shell/Esso			October 1974	Under appraisal
Unnamed	9/8	Hamilton			July 1974	Under appraisal

There are considerable quantities of associated gas with the oil in many of the East Shetland Basin fields.

**Production/Reserves figures and/or development plans not fully confirmed.*

APPENDIX 5

Estimated United Kingdom Continental Shelf Oil Reserves

	Totals (millions of tons)				
	Proven	Probable	Probable Total	Possible	Possible Total
1 Proven fields	995 (895)	90 (165)	1085 (1060)	135 (100)	1220 (1160)
2 Other significant discoveries not yet fully appraised	65 (–)	215 (230)	280 (230)	300 (160)	580 (390)
3 Total from existing finds (1 March 1975)	1060 (895)	305 (395)	1365 (1290)	435 (260)	1800 (1550)
4 Expected from future finds on existing licences	– (–)	900 (700)	900 (700)	400 (700)	1300 (1400)
5 Total from existing licences	1060 (895)	1205 (1095)	2265 (1990)	835 (960)	3100 (2950)

NOTES

Proven – those which on the available evidence are virtually certain to be technically and economically producible.

Probable – those which are estimated to have a better than 50 per cent chance of being technically and economically producible.

Possible – those which at present are estimated to have less than a 50 per cent chance of being producible.

The figures in brackets are the 1974 estimates.

(Reproduced from *Development of the Oil and Gas Resources of the UK*, 1975, by courtesy of HMSO, London.)

APPENDIX 6

Forecast Range of UK Oil Production 1975–1985

Production from existing and future discoveries in the presently designated areas of the United Kingdom Continental Shelf

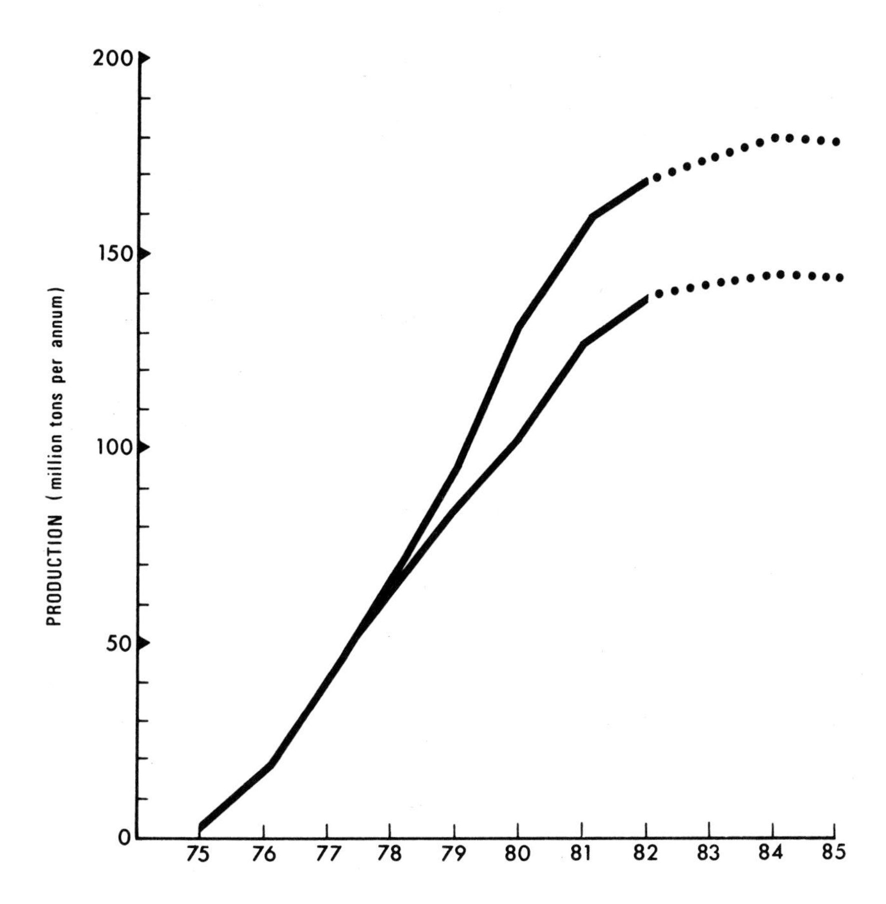

(Reproduced from *Development of the Oil and Gas Resources of the UK*, 1975, by courtesy of HMSO, London.)

APPENDIX 7

Estimated UK North Sea Gas Reserves *(as at 31 December 1974)*

	Totals in trillion (10^{12}) cubic feet			
	Proven	Probable	Possible	Total
Southern Basin				
Fields presently being produced or under contract to British Gas	18.2	1.1	1.5	20.8
Other discoveries believed to be commercial but not yet covered by British Gas contract	2.8	0.1	0.2	3.1
Other discoveries which may become commercial in due course	–	1.2	1.4	2.6
Total Southern Basin	21.0	2.4	3.1	26.5
Northern Basin				
Under contract to British Gas	2.9	0.3	–	3.2
Other significant gas discoveries (including gas-condensate finds)	–	4.3	4.5	8.8
Gas associated with oil discoveries	3.0	2.5	0.4	5.9
Total Northern Basin	5.9	7.1	4.9	17.9
TOTAL UK North Sea	26.9	9.5	8.0	44.4

(Reproduced from *Development of the Oil and Gas Resources of the UK*, 1975, by courtesy of HMSO, London.)

APPENDIX 8

Oil Production Platforms — May 1975

Field	Operator		Platform Contractor	Site	Platform type	
Platforms Installed						*Installation date*
Forties	BP	I	Laing Offshire	Teesside	steel	June 1974
Auk	Shell		Redpath Dorman Long	Methil	steel	July 1974
Forties	BP	II	Highland Fabricators	Nigg Bay	steel	August 1974
Argyll	Hamilton		Conversion by Wilson Walton	Teesside	converted drilling rig	March 1975
Platforms under construction						*Scheduled platform delivery date*
Piper	Occidental		McDermott Union Industrielle et d'Enterprise	Ardesier Le Havre, France	steel	1975
Brent	Shell	A	Redpath Dorman Long	Methil	steel	1975
		B	Hoyer Ellefson/Aker/Selmer	Stavanger, Norway	concrete	1975
Beryl	Mobil		Hoyer Ellefson/Aker/Selmer	Stavanger, Norway	concrete	1975
Montrose	Amoco		Union Industrielle et d'Enterprise	Le Havre, France	steel	1975
Forties	BP	III	Highland Fabricators	Nigg Bay	steel	1975
		IV	Laing Offshire	Teesside	steel	1975
Dunlin	Shell		Andoc	Rotterdam, Holland	concrete	1976
Thistle	Burmah		Laing Offshore	Teesside	steel	1976
Claymore	Occidental		Union Industrielle et d'Enterprise	Le Havre, France	steel	1976
Cormorant	Shell		McAlpine	Ardyne Point	concrete	1976
Brent	Shell	C	McAlpine	Ardyne Point	concrete	1976
		D	Hoyer Ellefson/Aker/Selmer	Stavanger, Norway	concrete	1976
Ninian	Chevron	I	Howard/Doris	Loch Kishorn	concrete	1977
		II	Highland Fabricators	Nigg Bay	steel	1977
Heather	Unocal		McDermott	Ardesier	steel	1977

(Reproduced from *Development of the Oil and Gas Resources of the UK*, 1975, by courtesy of HMSO, London)

APPENDIX 9

Equipment for a typical northern North Sea offshore oil production platform

The following list covers the purchases for one platform and is presented generically by type of equipment.

Description	Values (£)
Substructure	16,000,000
Material for Deck Modules	395,000
Living Quarters and Helicopter Deck	800,000
Gas/Oil Separators and Gas Scrubbers	180,200
Pumps	245,000
Air Compressors, Receivers and Driers	13,500
Boilers and Accessories	12,400
Burner Equipment	6,500
Water and Sewage Treating Plants	130,000
Tanks	8,000
Mixing Equipment	375
Pipe and Tube	83,000
Valves	148,550
Fittings	20,000
Flanges	23,000
Manifolds	10,100
Hydraulic Pipe Clamps	18,000
Bolting Material	4,100
Gaskets	1,500
Power Generation Units and Electrical Equipment	1,328,000
Power and Control Cable	48,000
Instrument, Meters and Gauges	95,000
Heating and Ventilating Equipment	13,900
Hoisting and Lifting Equipment	137,000
Communication Equipment	4,750
Fire Fighting and Safety Equipment	214,000
Navigational Aids	8,000
Wellhead Equipment (for 10 wells)	165,000
Hydraulic Shut-down Unit	30,000
Total:	**£20,142,875**

Prices as at November 1973

(Reproduced from *The pattern of capital purchases for a typical northern North Sea offshore oil production platform* by courtesy of Shell UK Exploration and Production Ltd)

APPENDIX 10

Oil Company Purchasing Offices — placing orders for goods and services

UK SECTOR

Amoco (UK) Exploration Co.,
St Alban's House, 59 Haymarket, London SW1Y 4QX.
For consumables: 25 Guild Street, Aberdeen AB1 2NJ.

BP Trading Ltd.,
BP House, Third Avenue, Harlow, Essex.
For consumables: BP Developments, PO Box 35, Shed 16, Queen Elizabeth Wharf, Dundee.

Burmah Oil (North Sea) Ltd.,
Salisbury House, London Wall, London EC2M 5XQ.
For consumables: 81-85 Waterloo Quay, Aberdeen AB2 1DE

Conoco Europe Ltd.,
Park House, 118 Park Street, London W1Y 4NN.
For consumables: Princess Alexandra Wharf, Dundee.

Hamilton Brothers Oil Company (Great Britain) Ltd.,
Cleveland House, 19 St James's Street, PO Box 17, London SW1Y 4LP.
For consumables: 6 Rubislaw Terrace, Aberdeen AB1 1XE.

Mobil Producing Northwest Europe Inc.,
Mobil House, 54-60 Victoria Street, London SW1E 6QB.
For consumables: Mobil North Sea, No 15 Bond, Abbotswell Road, Aberdeen AB1 4AD.

Occidental of Britain Inc.,
2nd Floor, 4 Grosvenor Place, London SW1.
For consumables: 39 Dee Street, Aberdeen AB1 1XE.

Phillips Petroleum Co.,
Portland House, Stag Place, London SW1E 5DA.
For consumables: Woodside Road, Bridge of Don, Aberdeen AB9 4AG.

Shell UK Exploration and Production Ltd.,
Shell Centre, London SE1 7NA.
For consumables: 1 Altens Farm Road, Aberdeen AB1 2PR.

Signal Oil and Gas Co. Ltd.,
173-176 Sloane Street, London SW1X 9QG.
For consumables: The Nord Centre, York Place, Waterloo Quay, Aberdeen AB9 2HA.

Texaco North Sea UK Company, 1 Knightsbridge Green, London
1 Knightsbridge Green, London SW1X 7QJ.
For consumables: Texaco Ltd., Charlotte House, Queen Street, Glasgow C1.

Total Oil Marine Ltd.,
Berkeley Square House, Berkeley Square, London W1X 6LT.
For consumables: Stell Road, Aberdeen AB1 2QR.

Unionoil Company of Great Britain,
32 Cadbury Road, Sunbury-on-Thames, Middlesex.
For consumables: 5 Regent Road, Aberdeen.

OTHER SECTORS

Elf Oil Exploration and Production (UK) Ltd.,
Knightsbridge House, 197 Knightsbridge, London SW7 1RZ.

Phillips Petroleum Co.,
Portland House, Stag Place, London SW1E 5DA.

Marathon Petroleum Ireland Ltd.,
Canada House, 65-68 St Stephen's Green, Dublin 2, Republic of Ireland.

Placid International Oil Ltd.,
Koningin Julianaplein 15, The Hague, The Netherlands.

Danish Underground Consortium,
The Gulf Oil Co. of Denmark, Lendermaerket, 11,
DK-999 Copenhagen, Denmark.

APPENDIX 11

What is the price of oil?

It depends on where it comes from, where it goes, its quality, what is done to refine it, and how it is taxed. Until recent changes the system was:

Posted price of 'market' crude: the price Opec increased so sharply from $2.47 a barrel on December 31 1972 to $11.65 on January 1 1974 (now $11.24, but see below). The 'market' grade is Saudi Arabian light crude, posted prices of other oils are geared to this. Nobody pays the posted price – it is the figure used to calculate income tax and royalty payments on equity crude oil.

Equity crude: oil belonging to the companies on which the tax and royalty payments are made. On an $11.25 posted price, comes to $9.76 – but add to this the 16 cents it costs to produce the stuff.

Participation crude: oil belonging to the host government (in proportion to its stake in the company). Mainly sold for distribution by the companies at a 'buy-back' price, geared to posted prices, of between $10.46 and $10.67.

Average price: what the cocktail of equity and participation oil average out to – $10.36 for a mix of 40 per cent equity and 60 per cent participation. (Note the drop in the posted price from $11.65 to $11.25 was offset by higher tax rates.)

Import cost, fob: average price plus profits and capital charges up to this stage. What oil cost and importing country's trade balance 'free-on-board'.

Delivered price: import price, cif, includes carriage, freight and insurance, which adds another $1 to Saudi Arabia crude delivered to Britain.

Product prices: now add refining costs, distributing costs and profits.

Consumer prices: the grand total including taxes levied by the consuming country's government.

Unitary prices: Opec has now officially adopted a unitary (or single) price system thus scrapping the nomenclature of posted prices, equity and participation oil. This gives a notional government revenue on market crude of $10.12 a barrel, but has yet to come fully into force.

Continued

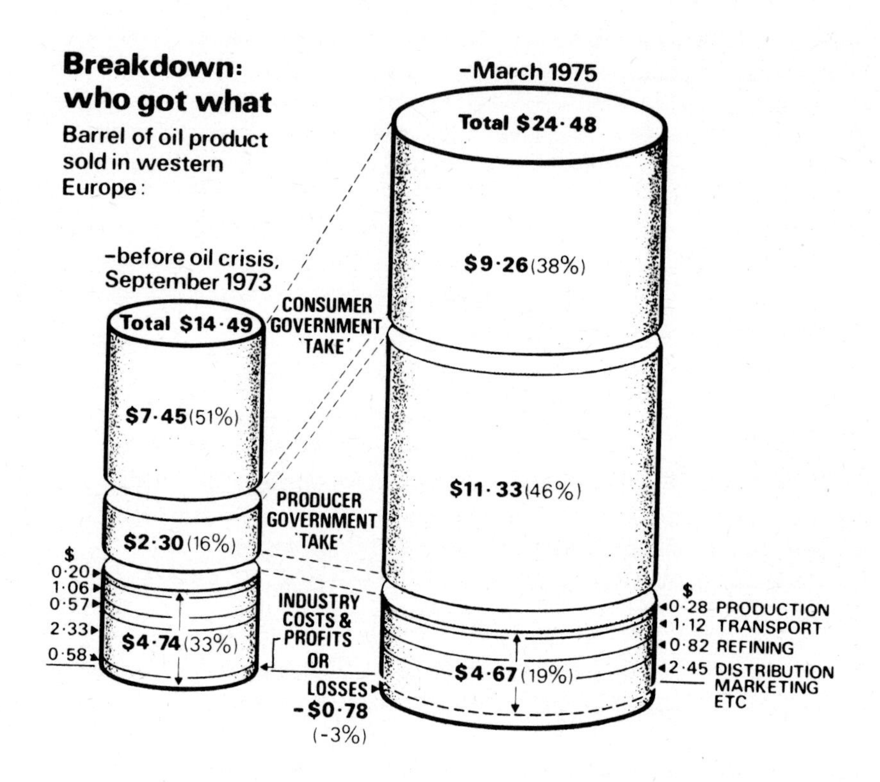

(Reproduced by courtesy of *The Economist*, April 26, 1975 issue.)

APPENDIX 12: Analysis of deliveries into consumption — gas, diesel and fuel oils — UK trade (excl. Derv)

(in tons)

INDUSTRY OR END USE	1973		1974	
	Gas, diesel oils	Fuel oils	Gas, diesel oils	Fuel oils
AGRICULTURE AND FORESTRY:				
Power Units	891,220	43,790	747,120	40,510
Driers and Heaters	334,630	324,850	262,470	265,180
MARINE:				
Fishing	344,880	162,210	337,570	130,750
Other Coastal and Inland Shipping	417,418	73,840	572,040	96,610
FOOD:				
Grain Milling	23,380	71,640	23,660	68,220
Baking	119,170	140,070	113,840	106,490
Milk Products	20,200	300,290	26,070	256,640
Sugar and Sugar Confectionery	23,640	327,320	14,370	300,120
Drink	77,240	754,310	76,150	765,660
Tobacco	10,410	49,830	9,490	42,130
Other	154,590	618,010	158,140	574,910
MINES AND QUARRIES	470,540	133,480	406,110	70,460
CHEMICALS:				
Soaps and Fats	5,150	232,560	4,220	224,770
Plastics	40,870	340,620	34,790	325,750
Other (incl. Petroleum Chemicals)	237,110	2,962,910	207,420	2,495,950

(Continued)

APPENDIX 12 (continued)

(in tons)

INDUSTRY OR END USE	1973		1974	
	Gas, diesel oils	Fuel oils	Gas, diesel oils	Fuel oils
METALS:				
Steel	438,300	4,294,850	371,770	3,438,560
Iron Castings	65,450	55,650	53,890	54,430
Non-Ferrous Metals	169,730	305,860	154,050	294,880
ENGINEERING:				
General – Non-Electrical	537,110	778,650	515,290	740,200
Electrical	170,040	478,640	157,660	418,960
Shipbuilding and Marine Engineering	67,340	77,190	58,850	56,450
Motor and Cycle Manufacture	154,750	371,840	148,120	339,250
Aircraft Manufacture	39,980	182,420	39,640	165,550
Other Vehicle Manufacture	45,010	47,300	40,270	32,050
Other Metal Manufacture	152,350	230,960	126,290	200,190
TEXTILES AND LEATHER:				
Man-made Fibres	38,930	528,620	37,660	456,360
Cotton	17,140	125,610	13,140	127,400
Wool	18,170	132,190	15,040	105,290
Other Textiles	66,640	429,630	61,930	371,830
Leather	14,730	66,630	11,150	61,540
Clothing	64,680	126,080	59,500	117,650

(Continued)

APPENDIX 12 (continued)

(in tons)

INDUSTRY OR END USE	1973		1974	
	Gas, diesel oils	Fuel oils	Gas, diesel oils	Fuel oils
BRICKS AND CERAMICS:				
Bricks and Other Building Materials	316,150	359,230	272,160	331,950
Pottery	14,430	23,160	12,250	21,660
Glass	37,700	820,580	35,080	741,610
Cement	32,200	402,080	29,240	411,200
TIMBER, RUBBER AND PAPER:				
Timber	59,910	61,370	53,720	50,430
Paper-making	49,070	1,304,900	35,270	1,176,660
Printing	72,690	139,200	75,850	123,190
Rubber Goods	26,440	196,330	22,510	150,820
OTHER MANUFACTURING INDUSTRIES	422,310	622,850	364,710	706,960
BUILDING AND CONTRACTING (incl. Open Cast Mining)	1,152,150	106,430	814,750	77,860
PUBLIC UTILITIES:				
Gas-Making	40,530	167,750	29,200	122,100
Electricity Generation	839,520	15,884,251	715,300	16,759,600
Water Supply	29,840	2,100	29,480	1,950
Railways	954,720	50,030	892,770	44,260
LAUNDRIES	67,230	177,590	58,970	156,240

(Continued)

APPENDIX 12 (continued)

(in tons)

INDUSTRY OR END USE	1973		1974	
	Gas, diesel oils	Fuel oils	Gas, diesel oils	Fuel oils
CENTRAL HEATING – Non-Industrial:				
Private Houses	663,720	13,280	590,200	23,900
Other Dwellings	224,430	53,560	216,730	40,510
Offices	385,170	143,250	354,980	144,360
Distributive Trades	469,050	307,120	429,310	262,960
Educational Establishments	882,820	262,600	818,250	190,880
Medical and Welfare Establishments	371,600	934,450	348,620	830,100
Religious Premises	217,220	7,440	177,630	5,700
Places of Entertainment	147,510	58,760	127,140	42,430
Catering Establishments	256,300	78,340	219,050	62,140
National Government Buildings	409,840	392,660	382,610	359,520
Local Government Buildings	378,140	182,370	401,170	141,950
British Armed Forces	224,790	295,010	205,700	229,330
Foreign Armed Forces	43,560	12,490	16,190	7,020
Other Premises	166,170	73,700	158,930	51,690
MISCELLANEOUS NON-MANUFACTURING	559,042	539,583	489,110	397,200
PETROLEUM INDUSTRY:				
Refinery Consumption	–	6,941,342	–	6,836,319
Other	116,090	383,340	131,710	324,280
TOTAL DELIVERIES INTO CONSUMPTION	**14,861,140**	**45,764,996**	**13,366,300**	**43,571,519**

(Reproduced by courtesy of the Institute of Petroleum)

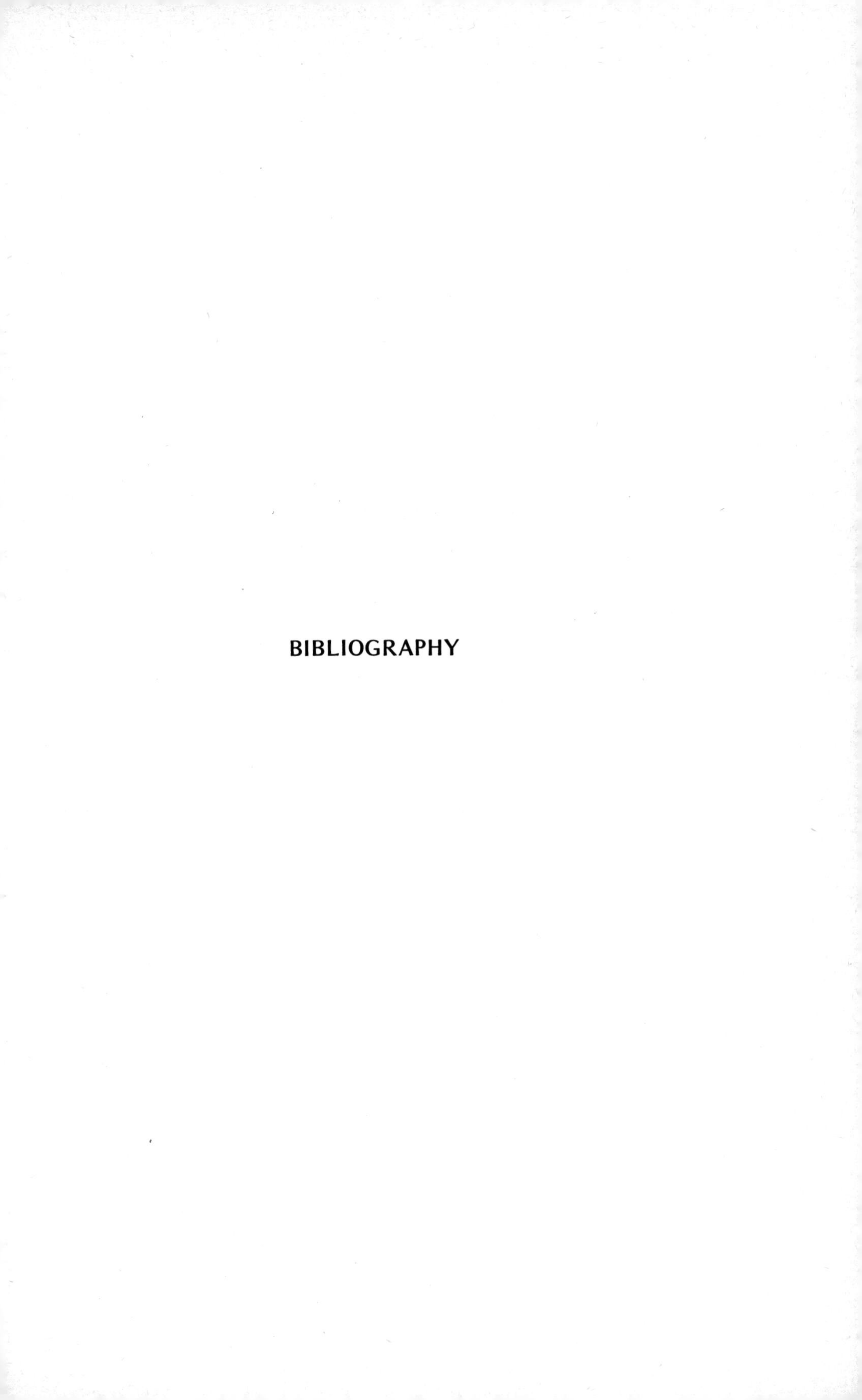

BIBLIOGRAPHY

BIBLIOGRAPHY

Books

BRANTLY, J.E., *Rotary Drilling Handbook,* Palmer Publications, London.

BURRELL, R.M., *Politics, Oil and the Western Mediterranean,* Sage Publications, London, 1973.

CALLOW, C., *Power from the Sea: the search for North Sea Oil and Gas,* Gollancz, London, 1973.

CAZENOVE & CO., *The North Sea, the search for oil and gas and the implications for investment,* London, 1972.

CBMPE, *British Petroleum Equipment and Services,* London.

CLARKSON, H. & CO., *The Tanker Register, Liquid Gas Carrier Register, Offshore Drilling Register,* London, 1975.

COLE, F.W. and MOORE, P.L., *Drilling Operations Manual,* Petroleum Publishing Company, 1965.

COOPER, B. and GASKELL, T.F., *North Sea Oil – The Great Gamble,* Heinemann, London, 1967.

CORNISH INSTITUTE OF ENGINEERS and CORNWALL COUNTY COUNCIL, *Oil Quest Cornwall, Cornwall and the Offshore Oil Industry,* County Planning Office, County Hall, Treyew Road, Truro, Cornwall.

COUNCIL OF BRITISH MANUFACTURERS OF PETROLEUM, *British Petroleum Equipment and Services Directory,* 118 Southwark Street, London SE1 0SU.

ECONOMISTS INTELLIGENCE UNIT, *Soviet Oil to 1980.*

FINANCIAL TIMES, *Oil and Gas International Year Book,* Business Enterprises Division, 10 Bolt Court, London EC4A 3HL.

FINANCIAL TIMES, *Financial Times, Who's Who in Oil and Gas,* Business Enterprises Division, 10 Bolt Court, London EC4A 3HL.

FRICK, T.C., *Petroleum Production Handbook, Vols I & II*, McGraw Hill, 1962.

GATLIN, C., *Petroleum Engineering*, Prentice Hall, 1968.

GRAY, A.W. and COLE, F.W., *Oil Well Drilling Technology*, University of Oklahoma Press, 1959.

GULF PUBLISHING CO. LTD., *Composite Catalogue of Oilfield Equipment and Services*, PO Box 2608, Houston, Texas 77001, USA; or from 131 Clapton Common, London E5 9AB.

HIGHLANDS & ISLANDS DEVELOPMENT BOARD, *Offshore Directory*, Bridge House, 27 Bank Street, Inverness.

HINDE, P., *The Exploration for Petroleum, with particular reference to North West Europe*, British Gas Corporation, 1974.

HOBSON, G.D. and POHI, W. (Eds), *Modern Petroleum Technology*, published by Applied Science Publishers, Ripple Road, Barking, Essex, on behalf of the Institute of Petroleum.

HOLS, A., *How to evaluate the Economics of North Sea Oilfields*, First International Offshore North Sea Technology Conference, Stavanger, September 1974. Rogaland Trade Fair Limited, Stavanger, Norway.

INSTITUTE OF OFFSHORE ENGINEERING, *Guide to information services in Marine Technology*, Heriot-Watt University, Chambers Street, Edinburgh EH1 1HX.

INSTITUTE OF PETROLEUM, *The Exploration of Petroleum in Europe and North Africa;* proceedings of a joint meeting between the Institute of Petroleum and the American Association of Petroleum Geologists, London 1969.

KOGAN PAGE LIMITED, *United Kingdom Offshore Oil and Gas Yearbook*, 116a Pentonville Road, London N1.

LOM, W.L. and WILLIAMS, A.F., *Liquefied Petroleum Gases*, Ellis Horwood Limited, London 1975.

MOODY, G.B., *Petroleum Exploration Handbook*, McGraw Hill, 1961.

MOSLEY, L., *Power-Play: the tumultuous world of Middle East Oil 1890–1973*, Weidenfeld and Nicholson, London 1973.

NORTH EAST SCOTLAND DEVELOPMENT AUTHORITY, *North East Scotland and the offshore oil industry*, Aberdeen 1973.

NORTH OF ENGLAND DEVELOPMENT COUNCIL and THE NORTHERN OFFSHORE SUPPLIES OFFICE, *Oilfield, a directory of offshore requirements and of industrial resources in the North of England.* Department of Industry, Higham House, New Bridge Street, Newcastle-upon-Tyne NE1 8AL.

NORTH SEA OIL DIRECTORY LTD, *North Sea Oil Directory,* 18 James Street, London W1M 5HN.

ODELL, P.R., *Oil and World Power: a geographical interpretation,* Penguin, London 1973.

OECD, *Oil: the present situation and the future prospects.* OECD/Paris, HMSO 1973.

PETROLEUM PUBLISHING CO., *MSH Oil Industry Directory,* PO Box 1260, Tulsa, Oklahoma 74101, USA, and London.

PETROLEUM PUBLISHING CO., *Offshore Contractors and Equipment Directory,* Tulsa, Oklahoma, and London.

PETROLEUM PUBLISHING CO., *Petroleum Directory – Eastern Hemisphere,* Tulsa, Oklahoma, or 17 Old Queen Street, London SW1.

PETROLEUM TIMES, *Guide to British Offshore Suppliers,* (in co-operation with the Offshore Supplies Office of the Department of Energy). IPC Industrial Press Ltd., 33-40 Bowling Green Lane, London EC1R 0NE.

POSNER, M., *Fuel Policy, a Study in Applied Economics,* Macmillan, London 1973.

SCOTTISH DEVELOPMENT DEPARTMENT, *North Sea Oil Production Platform Towers: Construction Sites* – a discussion paper, Edinburgh 1973.

SHIPSTATS, *Register of Offshore Supply Vessels,* Shipstats Subscription Dept, 49 St Mary's Road, Gillingham, Kent ME7 1JJ.

SKINNER, W.R., *Who's Who in World Oil and Gas,* Financial Times, London 1975.

SKINNER, W.R., *Skinner's Oil and Gas International Year Book,* Financial Times, London 1975.

SPEARHEAD PUBLICATIONS LTD, *Underwater Engineering Directory,* 2 Fife Road, Kingston-on-Thames, Surrey.

UNIVERSITY OF OKLAHOMA, SCIENCE AND PUBLIC POLICY PROGRAM, *Energy under the Oceans,* University of Oklahoma/Bailey Brothers and Swinfen, Oklahoma, 1973.

WALMESLEY, P.J., *North Sea Oil: the challenge and implications*, Heriot-Watt University, Edinburgh 1973.

WORLD OIL GULF PUBLISHING COMPANY, *Composite Catalogue of Oil Field Equipment and Services*, Houston, Texas.

WORLD OIL GULF PUBLISHING COMPANY, *Hydrocarbon Processing Catalogue – a pre-filed catalogue of equipment, materials and services, for processing crude oil, natural gas and petrochemicals*, Houston, Texas.

Journals

Anti Corrosion, Sawell Publications, London.

BSI NEWS, BSI Sales and Subscriptions, London.

Bulletin, Association for Petroleum and Explosives Administration, Huntingdon.

Chartered Mechanical Engineer, Institute of Mechanical Engineers, London.

Chemical Age, Morgan Grampian (Publishers) Ltd., London.

Compressed Air, Ingersoll Rand Co. Ltd., London.

Corrosion, National Association of Corrosion Engineers Inc., South Houston, Texas.

Drilling, Ingersoll Rand Co. Ltd., London.

Electrical Times, IPC Electrical-Electronic Press Ltd, London.

Energy Trends, Department of Energy, Room 1203, Thames House South, London SW1.

Hydrocarbon Processing, Gulf Publishing Co., Houston, Texas, and London.

Journal of the Institute of Petroleum, London.

Journal of Petroleum Technology, Society of Petroleum Engineers of AIME, Dallas, Texas.

Materials Performance, National Association of Corrosion Engineers Inc., Houston, Texas.

Material Research, National Association of Corrosion Engineers Inc., Houston, Texas.

Mechanical Engineering, American Society of Mechanical Engineers, New York.

Middle East Economic Digest, MEED, London.

Noroil, Noroil Publishing House Ltd., Aberdeen.

Northern Oil Digest, Aberdeen Journals Ltd., Lang Stracht, Mastrick, Aberdeen AB9 8AF.

North Sea Letter, The Financial Times Ltd., Subscription Department, Bracken House, 10 Cannon Street, London EC4P 4BY.

Ocean Industry, Gulf Publishing Co., Houston, Texas, and London.

Ocean Oil Weekly Report, Petroleum Publishing Co., Tulsa and London.

Offshore, Tulsa, Oklahoma.

Offshore Engineer, Thomas Telford Ltd, London.

Offshore Services, Spearhead Publications Ltd, Kingston-upon-Thames.

Oil Daily, 320 Campus Drive, Somerset, New Jersey 08873, USA.

Oil and Gas Journal, Petroleum Publishing Co., Tulsa, and London.

Oil and Gas Review, The Techress Group, London.

Oil News Service, Glasgow.

Oilman, Maclean Hunter Ltd., London.

Petroleum Economist, Petroleum Press Services, London.

Petroleum Engineer International, Petroleum Engineer Publishing Co., Dallas, Texas.

Petroleum Gazette, Petroleum Information Bureau, Melbourne, Australia.

Petroleum International, Petroleum Publishing Co., Tulsa and London.

Petroleum Review, The Institute of Petroleum, London.

Petroleum Times, IPC Industrial Press Ltd., London.

Pipeline, Pipeline News, Houston, Texas.

Pipeline and Gas Journal, Petroleum Engineer Publishing Co., Dallas, Texas.

Pipes and Pipeline International, Scientific Surveys Ltd, Beaconsfield, Buckinghamshire.

Process Engineering, Morgan Grampian Publishing Co., London.

Processing, IPC Industrial Press Ltd., London.

Power, McGraw Hill Publishing Co. Ltd, London.

Roustabout, Roustabout Publications, 35A Union Street, Aberdeen AB1 2BN.

Trade and Industry, HMSO, London.

Undersea Technology, Compass Publications Inc., Suite 1000, 1117N 19th Street, Arlington, Virginia, USA.

World Oil, Gulf Publishing Co., Houston and London.